The Fuzziness in Molecular, Supramolecular, and Systems Chemistry

The Fuzziness in Molecular, Supramolecular, and Systems Chemistry

Editor

Pier Luigi Gentili

MDPI • Basel • Beijing • Wuhan • Barcelona • Belgrade • Manchester • Tokyo • Cluj • Tianjin

Editor
Pier Luigi Gentili
University of Perugia
Italy

Editorial Office
MDPI
St. Alban-Anlage 66
4052 Basel, Switzerland

This is a reprint of articles from the Special Issue published online in the open access journal *Molecules* (ISSN 1420-3049) (available at: https://www.mdpi.com/journal/molecules/special_issues/Systems_Chemistry).

For citation purposes, cite each article independently as indicated on the article page online and as indicated below:

LastName, A.A.; LastName, B.B.; LastName, C.C. Article Title. *Journal Name* **Year**, *Article Number*, Page Range.

ISBN 978-3-03943-178-6 (Hbk)
ISBN 978-3-03943-179-3 (PDF)

Contents

About the Editor

Pier Luigi Gentili received his Ph.D. in Chemistry from the University of Perugia in 2004. His research and teaching activities are focused on Complex Systems. He is the author of the book "Untangling Complex Systems: A Grand Challenge for Science" published by CRC Press in 2018. Being aware that inanimate matter is driven by force-fields, whereas the interactions between biological systems are also information-based, Gentili is led by some questions like the following: When does a chemical system become intelligent? Is it possible to develop a "chemical artificial intelligence?" For the development of chemical artificial intelligence, Pier Luigi Gentili relies upon the theory and tools of Natural Computing. In particular, he is tracing a new path in the field of Neuromorphic Engineering by using non-linear chemical systems and by encoding information mainly through UV-visible signals. He is proposing methods to process fuzzy logic by molecular, supramolecular, and systems chemistry.

molecules

MDPI

Editorial

The Fuzziness in Molecular, Supramolecular, and Systems Chemistry

Pier Luigi Gentili

Department of Chemistry, Biology, and Biotechnology, Università degli Studi di Perugia, Via elce di sotto 8, 06123 Perugia, Italy; pierluigi.gentili@unipg.it; Tel.: +39-075-585-5573

Received: 6 August 2020; Accepted: 7 August 2020; Published: 10 August 2020

---

1. Introduction

The global challenges of the XXI century require a more in-depth analysis and investigation of complex systems [1]. A promising research line to better understand complex systems, and propose new algorithms and computing devices is natural computing. Natural computing is based on a fundamental rationale: every causal phenomenon can be conceived as a computation and every distinguishable physicochemical state of matter and energy can be used to encode information. Any physicochemical law can be exploited to make computations. For instance, quantum mechanics laws can be exploited to make quantum computing; the chemical kinetic laws can be used to make chemical computing; the laws of chaos to make chaos-computing, etc. On the other hand, we might draw inspiration from living beings with the exclusive attribute of using matter and energy to encode, collect, store, process, and send information [1,2]. Living beings show different information systems. Their basic information system is the cell, also called the biomolecular information system (BIS). In most multicellular organisms, we encounter nervous systems that constitute neural information systems (NISs). The defense systems that help repel antigens and disease-causing organisms are defined as immune information systems (IISs). Finally, most living beings live in societies, and the resulting aggregations constitute the so-called social information systems (SISs).

2. Artificial Intelligence and Fuzzy Logic

Among the natural information systems, particularly alluring for facing XXI century challenges, is the human nervous system (HNS). Its performances are astonishing. Based on a complex architecture of billions of nerve cells, our nervous system allows us to handle accurate and vague information by computing with numbers and words. Furthermore, it allows us to recognize variable patterns quite easily and make decisions in complex situations. Therefore, it is worthwhile trying to understand how it works and mimic it by developing artificial intelligence (AI). Within AI, fuzzy logic stands out as a good model of the human ability to compute with words and make decisions in complex circumstances [3,4]. Its descriptive and modeling power hinges on the structural and functional analogies it has with the HNS [5,6]. The entire architecture of the HNS is related to that of any fuzzy logic system. Our natural sensors play as fuzzifiers, our brain as a fuzzy inference engine, and our effectors as defuzzifiers. Every sensory system, physical and chemical, such as the visual or olfactory system, is constituted by a tissue of a spatially distributed array of sensory cells that behave as fuzzy sets [5,6]. Within each sensory cell, there is a multitude of sensory proteins that work as molecular fuzzy sets. The multiple information of any stimulus, i.e., its modality, intensity, spatial distribution, and time-evolution, is encoded hierarchically as degrees of membership to the molecular and cellular fuzzy sets. The imitation of these features allows us to design new artificial sensory systems with enhanced discriminative power due to different molecular fuzzy sets' parallel activity. A concrete example is the recent implementation of biologically inspired photochromic fuzzy logic systems that extend human vision to the UV [7,8].

3. Neuromorphic Engineering and Chemical Artificial Intelligence

The mimicry of nonlinear neural dynamics is a promising alternative strategy to approach human intelligence performances. Surrogates of neurons can be achieved through either oscillatory or excitable or chaotic chemical systems in solution (i.e., wetware) [9,10] or the solid phase (i.e., hardware) [11–13]. In this Special Issue, Szaciłowski and his team present an experimental characterization of an optoelectronic device, constituted by a polycrystalline cadmium sulfide electrode [14]. Such a device realizes a type of short-term plasticity, i.e., the paired-pulsed facilitation (PPF). The PPF consists of an enhancement in the postsynaptic current when the excitatory signal frequency increases. This short-term memory effect confers to the device an appreciable power of recognizing hand-written numbers. Szaciłowski's work blazes a trail for the optoelectronic implementation of neural network architectures that will allow the processing of fuzzy logic and recognition of variable patterns. Suffice to think that fuzzy logic has already been implemented through a pacemaker neuron model, such as the Belousov-Zhabotinsky reaction [15], and a chaotic neuron model, such as the "photochemical oscillator" [16,17]. When UV–visible radiation is chosen as a signal, it is straightforward to implement neuromodulation [18] and hence, fuzzy logic.

In the orthodox AI, fuzzy logic is processed through software running in digital electronic computers; it is even better if the electronic circuits are analog, since fuzzy logic is an infinite-valued logic. In the burgeoning field of chemical artificial intelligence (CAI) [19], unconventional chemical systems have been put forward for implementing fuzzy logic systems. Some examples can be found in the references [20–26]. The fundamental requirement is to have smooth analog input–output relationships between physicochemical variables, either linear or hyperbolic, but certainly not sigmoid. Sigmoid functions are adequate for processing discrete logics [27,28].

4. Cellular Fuzziness

The relentless investigation of the working principles of the cells or BISs has been revealing cellular fuzziness. Some proteins play within any cell as if they were the neurons of the "cellular nervous system" [6]. They are the protagonists of the signaling and genetic networks, and they make the cell capable of responding to ever-changing environmental conditions. As Fuxreiter points out in her perspective included in this Special Issue [29], often, a protein exists as a heterogeneous ensemble of conformers. For these proteins, the deterministic inference "amino-acidic sequence → 3D structure → function" is not applicable. In fact, the conformational ensemble may perform multiple functions, depending on the context. Such a collection of conformers looks like a macromolecular fuzzy set. The dynamical power of a protein to autonomously select a context-dependent function constitutes what we might name as its fuzzy inference engine. In their review, Jeffery and Liu tell us that there are moonlighting proteins in the metabolic network of a cell, in which one polypeptide chain performs more than one physiologically relevant biochemical or biophysical function [30]. The kind of function that is executed might depend on cellular localization, concentrations of substrates or ligands, or environmental stress. Any type of moonlighting protein is fuzzy because some of its copies can perform one function, some another, and some both functions simultaneously. As cellular conditions change due to metabolism and environmental conditions, the functions of these proteins change as well. Uversky informs us that within a cell, the supramolecular interactions between specific intrinsically disordered proteins and hybrid proteins, having ordered domains and intrinsically disordered protein regions, drive biological liquid–liquid phase transitions that form proteinaceous membrane-less organelles (PMLOs) [31]. PMLOs are intracellular hot spots that serve as organizers of cellular biochemistry. Such PMLOs are fuzzy, and their fuzziness resides in their compositional and compartmental variety and variability. Dodero and her team made the tangible experience of supramolecular fuzziness by investigating the interaction between a Transcription Factor and double-stranded DNA [32]. After annealing a proper DNA sequence and synthesizing a photosensitive surrogate of the GCN4 Transcription Factor, Dodero and her colleagues furnish experimental evidence of the protein-DNA complexation fuzziness by using different techniques, such as NMR, electrophoretic mobility shift assay, and circular dichroism spectroscopy. To monitor the conformational fuzziness of macromolecules and smaller molecules, Gentili relies upon the maximum entropy method to extract the distributions of conformers from any kinetic trace [6,33]. After determining the distribution, quantifying its fuzzy entropy is also possible [6,34].

5. Non-Arrhenius Kinetics

If we consider the conformational distributions of compounds, the original transition-state theory and the Arrhenius law might appear far-fetched. There is a peculiar distribution of conformers at every temperature, and every conformer traces its unique reactive path. It is not fair to define just one kinetic constant and one activation energy for all the coexistent conformers. It is necessary to add that both the original transition-state theory and the Arrhenius law have been already questioned by the most recent theoretical and experimental developments, as evidenced by Carvalho-Silva, Coutinho, and Aquilanti [35]. Quantum mechanical effects, such as tunneling and resonance, stochastic motions of particles in condensed environments, and non-equilibrium effects in classical and quantum formulations, are responsible for deviations from the traditional Arrhenius equation. In such situations, the transitivity function, defined in terms of the reciprocal of the apparent activation energy, measures the propensity for a reaction to proceed. The transitivity function provides a tool for implementing phenomenological kinetic models. In reference [36], Machado, Sanches-Neto, Coutinho, Mundim, Palazzetti, and Carvalho-Silva document the general scope of a transitivity code that can estimate the kinetic and thermodynamic parameters of physicochemical processes and deal with non-Arrhenius behavior.

6. Conclusions and Perspectives

This Special Issue's multidisciplinary contributions highlight that the theory of fuzzy set and fuzzy logic are valuable conceptual tools to understand the molecular and supramolecular world. Of course, quantum-mechanics already exists for this purpose, but fuzzy logic is becoming an alternative approach that might have still undiscovered common points with quantum logic [37,38]. Fuzzy logic appears particularly suitable for dealing with conformers. Although this approach is in its infancy, it is worthwhile pursuing it. It will allow us to describe any cell's activities, the constitutive elements of the human nervous system, and the immune system's performances more deeply. Such knowledge will be translated into new strategies to control the cellular processes and develop chemical artificial intelligence and chemical robots [6]. If cutting-edge technologies emerge from this approach, then, biomolecular, supramolecular, and systems chemistry will surely be considered fuzzy worldwide!

Funding: This research was funded by ANVUR grant number n.20/2017.

Acknowledgments: P.L. Gentili acknowledges all the contributors to this Special Issue, the anonymous reviewers of the papers published within this Special Issue, and Lola Huo for her valuable editorial assistance.

Conflicts of Interest: The author declares no conflict of interest.

References

1. Gentili, P.L. *Untangling Complex Systems: A Grand Challenge for Science*, 1st ed.; CRC Press, Taylor & Francis Group: Boca Raton, FL, USA, 2018.
2. Cronin, L.; Walker, S.I. Beyond prebiotic chemistry. *Science* **2016**, *352*, 1174–1175. [CrossRef] [PubMed]
3. Zadeh, L.A. Fuzzy logic-a personal perspective. *Fuzzy Sets Syst.* **2015**, *281*, 4–20. [CrossRef]
4. Zadeh, L.A. Outline of a new approach to the analysis of complex systems and decision processes. *IEEE Trans. Syst. Man Cybern.* **1973**, *3*, 28–44. [CrossRef]
5. Gentili, P.L. The human sensory system as a collection of specialized fuzzifiers: A conceptual framework to inspire new artificial intelligent systems computing with words. *J. Intell. Fuzzy Syst.* **2014**, *27*, 2137–2151. [CrossRef]
6. Gentili, P.L. The Fuzziness of the Molecular World and Its Perspectives. *Molecules* **2018**, 23, 2074. [CrossRef]

7. Gentili, P.L.; Rightler, A.L.; Heron, B.M.; Gabbutt, C.D. Extending human perception of electromagnetic radiation to the UV region through biologically inspired photochromic fuzzy logic (BIPFUL) systems. *Chem. Comm.* **2016**, *52*, 1474–1477. [CrossRef]
8. Gentili, P.L.; Rightler, A.L.; Heron, B.M.; Gabbutt, C.D. Discriminating between the UV-A, UV-B and UV-C regions by novel Biologically Inspired Photochromic Fuzzy Logic (BIPFUL) systems: A detailed comparative study. *Dyes Pigment.* **2016**, *135*, 169–176. [CrossRef]
9. Gentili, P.L.; Giubila, M.S.; Germani, R.; Romani, A.; Nicoziani, A.; Spalletti, A.; Heron, B.M. Optical Communication among Oscillatory Reactions and Photo-Excitable Systems: UV and Visible Radiation Can Synchronize Artificial Neuron Models. *Angew. Chem. Int. Ed.* **2017**, *56*, 7535–7540. [CrossRef]
10. Gentili, P.L.; Giubila, M.S.; Germani, R.; Heron, B.M. Photochromic and luminescent compounds as artificial neuron models. *Dyes Pigment.* **2018**, *156*, 149–159. [CrossRef]
11. Marković, D.; Mizrahi, A.; Querlioz, D.; Grollier, J. Physics for neuromorphic computing. *Nat. Rev. Phys.* **2020**. [CrossRef]
12. Lee, Y.; Lee, T.-W. Organic synapses for neuromorphic electronics: From brain-inspired computing to sensorimotor nervetronics. *Acc. Chem. Res.* **2019**, *52*, 964–974. [CrossRef] [PubMed]
13. Przyczyna, D.; Zawal, P.; Mazur, T.; Strzelecki, M.; Gentili, P.L.; Szaciłowski, K. In-materio neuromimetic devices: Dynamics, information processing and pattern recognition. *Jpn. J. Appl. Phys.* **2020**, *59*, 050504. [CrossRef]
14. Przyczyna, D.; Lis, M.; Pilarczyk, K.; Szaciłowski, K. Hardware Realization of the Pattern Recognition with an Artificial Neuromorphic Device Exhibiting a Short-Term Memory. *Molecules* **2019**, *24*, 2738. [CrossRef] [PubMed]
15. Gentili, P.L.; Horvath, V.; Vanag, V.K.; Epstein, I.R. Belousov-Zhabotinsky "chemical neuron" as a binary and fuzzy logic processor. *Int. J. Unconv. Comput.* **2012**, *8*, 177–192.
16. Gentili, P.L.; Dolnik, M.; Epstein, I.R. "Photochemical oscillator": Colored hydrodynamic oscillations and waves in a photochromic system. *J. Phys. Chem. C* **2014**, *118*, 598–608. [CrossRef]
17. Gentili, P.L.; Gotoda, H.; Dolnik, M.; Epstein, I.R. Analysis and prediction of aperiodic hydrodynamic oscillatory time series by feed-forward neural networks, fuzzy logic, and a local nonlinear predictor. *Chaos* **2015**, *25*, 013104. [CrossRef]
18. Bartolomei, B.; Heron, B.M.; Gentili, P.L. A contribution to neuromorphic engineering: Neuromodulation implemented through photochromic compounds maintained out of equilibrium by UV-visible radiation. *Rend. Fis. Acc. Lincei* **2020**, *31*, 39–52. [CrossRef]
19. Gentili, P.L. Small steps towards the development of chemical artificial intelligent systems. *RSC Adv.* **2013**, *3*, 25523–25549. [CrossRef]
20. Gentili, P.L. Boolean and fuzzy logic implemented at the molecular level. *Chem. Phys.* **2007**, *336*, 64–73. [CrossRef]
21. Gentili, P.L. Boolean and Fuzzy Logic Gates Based on the Interaction of Flindersine with Bovine Serum Albumin and Tryptophan. *J. Phys. Chem. A* **2008**, *112*, 11992–11997. [CrossRef]
22. Deaton, R.; Garzon, M. Fuzzy logic with biomolecules. *Soft Comput.* **2001**, *5*, 2–9. [CrossRef]
23. Zadegan, R.M.; Jepsen, M.D.E.; Hildebrandt, L.L.; Birkedal, V.; Kjems, J. Construction of a Fuzzy and Boolean Logic Gates Based on DNA. *Small* **2015**, *11*, 1811–1817. [CrossRef] [PubMed]
24. Gentili, P.L. The fundamental Fuzzy logic operators and some complex Boolean logic circuits implemented by the chromogenism of a spirooxazine. *Phys. Chem. Chem. Phys.* **2011**, *13*, 20335–20344. [CrossRef]
25. Pilarczyk, K.; Daly, B.; Podborska, A.; Kwolek, P.; Silverson, V.A.D.; de Silva, A.P.; Szaciłowski, K. Coordination chemistry for information acquisition and processing. *Coord. Chem. Rev.* **2016**, *325*, 135–160. [CrossRef]
26. Bhattacharjee, D.; Kim, W.; Chattopadhyay, A.; Waser, R.; Rane, V. Multi-valued and Fuzzy Logic Realization using TaOx Memristive Devices. *Sci. Rep.* **2018**, *8*, 1–10.
27. Arkin, A.; Ross, J. Computational functions in biochemical reaction networks. *Biophys. J.* **1994**, *67*, 560–578. [CrossRef]
28. Gentili, P.L. Molecular Processors: From Qubits to Fuzzy Logic. *ChemPhysChem* **2011**, *12*, 739–745. [CrossRef]
29. Fuxreiter, M. Towards a Stochastic Paradigm: From Fuzzy Ensembles to Cellular Functions. *Molecules* **2018**, *23*, 3008. [CrossRef]
30. Liu, H.; Jeffery, C.J. Moonlighting Proteins in the Fuzzy Logic of Cellular Metabolism. *Molecules* **2020**, *25*, 3440. [CrossRef]

31. Uversky, V.N. Supramolecular Fuzziness of Intracellular Liquid Droplets: Liquid–Liquid Phase Transitions, Membrane-Less Organelles, and Intrinsic Disorder. *Molecules* **2019**, *24*, 3265. [CrossRef]
32. Quirolo, Z.B.; Sequeira, M.A.; Martins, J.C.; Dodero, V.I. Sequence-Specific DNA Binding by Noncovalent Peptide–Azocyclodextrin Dimer Complex as a Suitable Model for Conformational Fuzziness. *Molecules* **2019**, *24*, 2508. [CrossRef] [PubMed]
33. Gentili, P.L. The fuzziness of a chromogenic spirooxazine. *Dyes Pigment.* **2014**, *110*, 235–248. [CrossRef]
34. Gambucci, M.; Gentili, P.L.; Sassi, P.; Latterini, L. A multi-spectroscopic approach to investigate the interactions between Gramicidin A and silver nanoparticles. *Soft Matter* **2019**, *15*, 6571–6580. [CrossRef] [PubMed]
35. Carvalho-Silva, V.H.; Coutinho, N.D.; Aquilanti, V. From the Kinetic Theory of Gases to the Kinetics of Rate Processes: On the Verge of the Thermodynamic and Kinetic Limits. *Molecules* **2020**, *25*, 2098. [CrossRef]
36. Machado, H.G.; Sanches-Neto, F.O.; Coutinho, N.D.; Mundim, K.C.; Palazzetti, F.; Carvalho-Silva, V.H. "Transitivity": A Code for Computing Kinetic and Related Parameters in Chemical Transformations and Transport Phenomena. *Molecules* **2019**, *24*, 3478. [CrossRef]
37. Rouvray, D.H. *Fuzzy Logic in Chemistry*, 1st ed.; Academic Press: San Diego, CA, USA, 1997.
38. Syropoulos, A. On vague chemistry. *Found. Chem.* **2020**. [CrossRef]

molecules

MDPI

Perspective

The Fuzziness of the Molecular World and Its Perspectives

Pier Luigi Gentili

Dipartimento di Chimica, Biologia e Biotecnologie, Università di Perugia, Via Elce di sotto 8, 06123 Perugia, Italy; pierluigi.gentili@unipg.it; Tel.: +39-075-585-5573

Received: 30 July 2018; Accepted: 17 August 2018; Published: 19 August 2018

Abstract: Scientists want to comprehend and control complex systems. Their success depends on the ability to face also the challenges of the corresponding computational complexity. A promising research line is artificial intelligence (AI). In AI, fuzzy logic plays a significant role because it is a suitable model of the human capability to compute with words, which is relevant when we make decisions in complex situations. The concept of fuzzy set pervades the natural information systems (NISs), such as living cells, the immune and the nervous systems. This paper describes the fuzziness of the NISs, in particular of the human nervous system. Moreover, it traces three pathways to process fuzzy logic by molecules and their assemblies. The fuzziness of the molecular world is useful for the development of the chemical artificial intelligence (CAI). CAI will help to face the challenges that regard both the natural and the computational complexity.

Keywords: fuzzy logic; complexity; chemical artificial intelligence; human nervous system; fuzzy proteins; conformations; photochromic compounds; qubit

1. Introduction

The scientific method, officially born in the seventieth century with the contributions of Galileo Galilei and Isaac Newton, has allowed humanity to become acquainted with the natural phenomena as never before. The acquisition of new scientific knowledge has also promoted an outstanding technological development in the last three hundred years or so. A mutual positive feedback relationship subsists between science and technology. To date amazing scientific and technological achievements have been reached. For example, we can explore the regions of the universe that are 10^{26} m far apart from us. At the same time, we can detect subatomic particles that have radii of the order of 10^{-15} m. We can record microscopic phenomena that occur in 10^{-18} s, but we can also retrieve traces of cosmic events happened billions of years ago. Our technology allows us to send robots to other planets of our solar system (e.g., the NASA Spirit rover on Mars), manipulate atoms and interfere with the expression of genes in living beings. Despite many efforts, there are still challenges that must be won. For instance, (I) we cannot predict catastrophic events on Earth (such as earthquakes and volcanic eruptions); (II) we strive to avoid the climate change; (III) we would like to exploit the energy and food resources without deteriorating the natural ecosystems and their biodiversity; (IV) there are diseases that are still incurable; (V) we would like to eradicate the poverty in the world; (VI) we make efforts to avoid or at least predict both economic and political crisis. Whenever we try to address such challenges, we experience frustrating insurmountable obstacles. Why? Because whenever we cope with one of them, we deal with a complex system. A complex system is one whose science is unable to give a complete and accurate description. In other words, scientists find difficulties in rationalizing and predicting the behaviors of complex systems. Examples of complex systems are the geology and the climate of the Earth; the ecosystems; each living being, in particular humans, giving rise to economic and social organizations, which are other examples of complex systems. The description of complex systems requires the collection, manipulation, and storage of big data [1], and the solution of problems of computational complexity. The description of complex systems from their ultimate constituents, i.e., atoms, is beyond our reach since the computational cost grows exponentially with the number of particles [2]. Moreover, many complex systems exhibit variable patterns. These variable patterns are objects (both inanimate and animate) or events whose recognition is made difficult by their multiple features, variability, and extreme sensitivity on the context. We still lack universally valid and effective algorithms for recognizing variable patterns [3]. Therefore, the obvious question is: How can we try to tackle the challenges regarding complex systems which involve issues of computational complexity? There are two principal strategies [4,5]. One consists in improving current electronic computers to make them faster and faster, and with increasingly large memory space. The other strategy is the interdisciplinary research line of natural computing. Researchers working on natural computing draw inspiration from Nature to propose: (I) new algorithms, (II) new materials and architecture for computing, and (III) new models to interpret complex systems. The sources of inspiration are the natural information systems, such as (a) the cells (i.e., the biomolecular information systems or BIS), (b) the nervous system (i.e., the neural information systems or NIS), (c) the immune system (i.e., the immune information systems or IIS), and (d) the societies (i.e., the societal information systems or SIS). Alternatively, we may exploit any causal event, involving inanimate matter, to make computation. In fact, in a causal event, the causes are the inputs and the effects are the outputs of a computation whose algorithm is defined by the laws governing the transformation (see Figure 1).

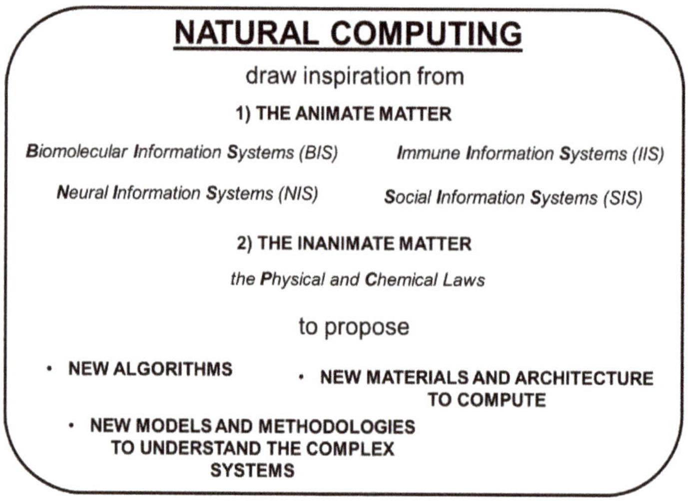

Figure 1. The contribution of the natural computing in coping with the challenges of the computational and natural complexity.

Among the natural information systems, the attention of many scientists worldwide is focused on the human nervous system that has human intelligence as its emergent property. The imitation of human intelligence is having a revolutionary impact in science, medicine, economy, security and well-being [6]. In fact, conventional quantitative techniques of system analysis are intrinsically unsuited for dealing with biological, social, economic, and any other type of system in which it is the behavior of the animate constituents that plays a dominant role. For such "humanistic systems", the principle of incompatibility holds [7]: as the complexity of a system increases, our ability to make accurate and yet significant statements about its behavior diminishes until a threshold is reached beyond which accuracy and significance (or relevance) become almost mutually exclusive characteristics. An alternative approach is based on the human intelligence that has the remarkable power of handling both accurate and vague information. Information is vague when it is based on sensory perceptions. Vague information is coded through the words of our natural languages. Therefore, humans compute by using not only numbers but also and especially words. We have the remarkable capability to reason, speak, discuss and make rational decisions without any quantitative measurement and any numerical computation, in an environment of uncertainty, partiality, and relativity of truth. Moreover, we recognize quite easily variable patterns, such as human faces and voices. Therefore, a major challenge of the artificial intelligence research line is the comprehension and implementation of the capabilities of the human intelligence to compute with words [8]. The use of classical, Aristotelian, divalent logic implemented in electronic circuits and computers has allowed reproducing and even overcoming the human ability to compute with numbers. The imitation of human ability to compute with words is still challenging. Fuzzy logic is a good model. In fact, fuzzy logic has been defined as a rigorous logic of the vague and approximate reasoning [9]. In this paper, after describing the principal features of fuzzy logic, it is demonstrated that one reason why fuzzy logic is a valid model of the human power to compute with words can be found at the molecular level. Therefore, we propose the use of molecular, supramolecular, and chemical systems as an innovative strategy for implementing fuzzy logic. This article wants to pursue the idea of developing a chemical artificial intelligence [10], i.e., an artificial intelligence that is based not on electronic circuits and software, but on chemical reactions in a wetware. Probably, the chemical artificial intelligence will promote the design of a new generation of computational machines, more similar to the brain rather than to the electronic computers. These new brain-like "chemical computers" should help to cope with the challenges regarding the complex systems, aforementioned in this Introduction.

2. Some Features of Fuzzy Logic

Fuzzy logic is based on the theory of fuzzy sets proposed by the engineer Lotfi Zadeh in 1965 [11]. A fuzzy set is different from a classical Boolean set. A classical set, also named as a crisp set, is a container that wholly includes or wholly excludes any given element. The theory of classical sets is based on the Law of Excluded Middle formulated by Aristotle in the fourth century BC. The Law of Excluded Middle states that an element x belongs to either set S or to its complement, i.e., set not-S. Zadeh proposed a refinement of the theory of the classical sets. In fact, a fuzzy set is more than a classical set: it can wholly include or wholly exclude elements, but it can also partially include and exclude other elements. The theory of fuzzy sets breaks the Law of Excluded Middle because an element x may belong to both set S and its complement not-S. An element x may belong to any set, but with different degrees of membership. The degree of membership (μ) of an element to a fuzzy set can be any real number included between 0 and 1. If $\mu = 0$, the element does not belong at all to the set; if $\mu = 1$, it completely belongs to the set; if $0 < \mu < 1$, the element belongs partially to the set. The Law of Excluded Middle is the foundation of the binary logic. In binary logic any variable is partitioned in two classical sets after fixing a threshold value: one set includes all the values below the threshold, whereas the other one contains those above. In the case of a positive logic convention, all the values of the first set become the binary 0, whereas those of the other set become the binary 1. The shape of a classical set is like that shown in Figure 2A. The degree of membership function for such a set discontinuously changes from 0 (below the threshold) to 1 (above the threshold). On the other hand, fuzzy sets can have different shapes. They can be sigmoidal, triangular, trapezoidal, Gaussian (see Figure 2), to cite a few. For a fuzzy set, the degree of membership function (μ) changes from 0 to 1. μ is the fuzzy unit of information, called "fit". It derives that fuzzy logic is an infinite-valued logic.

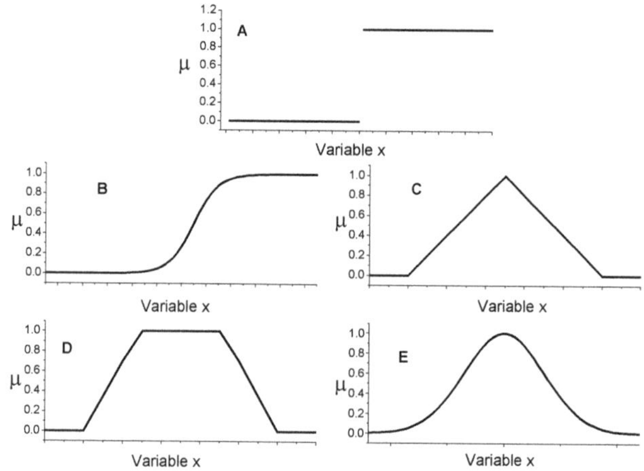

Figure 2. Shapes of the membership functions (μ) for a generic variable x: the case of a classical Boolean set in **A**; examples of fuzzy sets with sigmoidal, triangular, trapezoidal, and Gaussian shapes are shown in **B–E** plots, respectively.

Fuzzy logic can be used to describe any non-linear cause and effect relationship by building a fuzzy logic system (FLS). The construction of an FLS requires three fundamental steps. First, the granulation of all the variables in fuzzy sets. The number, position, and shape of the fuzzy sets are context-dependent. Second, the graduation of all the variables. A word, often an adjective, labels every fuzzy set. Third, the relationships between input and output fuzzy sets are described through syllogistic statements of the type "IF … , THEN … .", called fuzzy rules. The "IF … " part is the antecedent and involves the linguistic labels chosen for the input fuzzy sets. The "THEN … " part is the consequent and involves the linguistic labels chosen for the output fuzzy sets.

When we have multiple inputs, these are connected through the AND, OR, NOT operators [12]. AND corresponds to the intersection (e.g., the intersection of two fuzzy sets, whose membership functions are μ_{S_1} and μ_{S_2}, can be $\mu_{S_1 \cap S_2} = \min[\mu_{S_1}, \mu_{S_2}]$ or $\mu_{S_1 \cap S_2} = \mu_{S_1} \times \mu_{S_2}$); OR corresponds to the union (e.g., the union of the two sets S_1 and S_2 can be $\mu_{S_1 \cup S_2} = \max[\mu_{S_1}, \mu_{S_2}]$ or $\mu_{S_1 \cup S_2} = \mu_{S_1} + \mu_{S_2} - \mu_{S_1} \times \mu_{S_2}$); NOT corresponds to the complement (e.g., the membership function for the Fuzzy complement of S is $\mu_{\overline{S}} = 1 - \mu_S$). Fuzzy rules may be provided by experts or can be extracted from numerical data. After the granulation, the graduation of all the input and output variables, and the formulation of the fuzzy rules, we have a FLS that is a predictive tool or a decision support system for the particular phenomenon it describes. The way an FLS works is schematically depicted through an example in Figure 3.

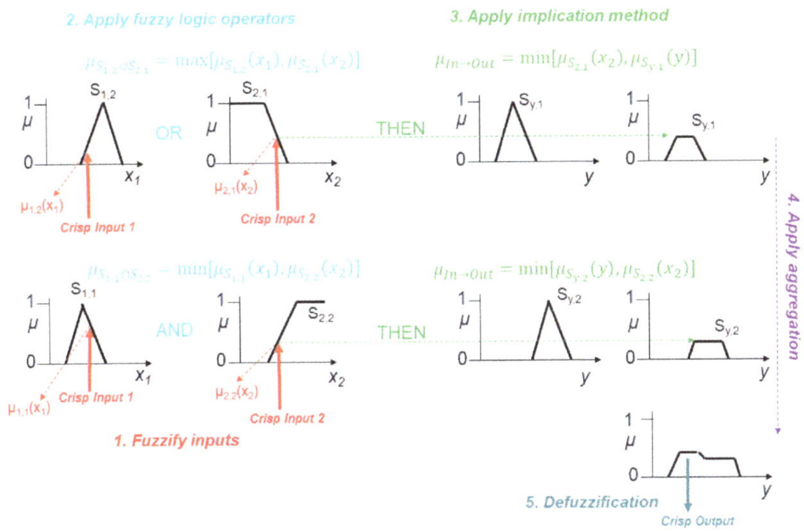

Figure 3. The flow of information in a fuzzy logic system where AND, OR and the implication have been implemented through the minimum, the maximum, and the minimum operators, respectively.

The information flows along the path traced by the arrows. First, the two crisp inputs are transformed in degrees of membership to the input fuzzy sets. This step is the so-called fuzzification process. It turns on all the fuzzy rules that involve the input Fuzzy sets "activated" by the crisp inputs. Second, the logic operators (AND, OR in Figure 3) combine the degrees of membership of the input fuzzy sets belonging to the two input variables. Third, the fuzzy implication method transforms the output fuzzy sets of each activated fuzzy rule through either the minimum or the product operator (in Figure 3, the minimum operator is used). Fourth, the activated output fuzzy sets are in turn aggregated through the maximum operator. Finally, the defuzzification procedure coverts the output Fuzzy sets in a crisp output value. The defuzzification method can be "the mean of the maxima", "the centroid", and others (for more information, see the tutorial by Mendel [12]). In a control-system application, the crisp output corresponds to a control action. In a signal processing application, such a number corresponds to a forecast or the location of a target. Fuzzy logic systems with adaptive capabilities are also used to predict chaotic time series [13,14]. The Fuzzy logic rules work as patches covering the chaotic attractors in their phase space. The rules are established through a learning procedure requiring a training data set.

The simulation and analysis of the dynamics of complex systems can be accomplished by the fuzzy cognitive maps (FCMs) [15]. The FCMs are an extension of the cognitive maps introduced by Axelrod [16]. An FCM is a graph, which consists of nodes and edges. The nodes represent concepts relevant to a given complex system, and edges represent the causal relationships among the nodes. Each edge is associated with a number that determines the degree of causal relation. The strengths of the relationships are usually normalized to the $[-1, +1]$ range. Value of -1 is full negative, +1 full positive, and 0 denotes no causal effect. The structure of an FCM is represented by a square matrix, called connection matrix, which reports all the weight values for edges between corresponding concepts represented by rows and columns. A complex system with n nodes will be represented by $n \times n$ connection matrix. The prediction of the evolution of a complex system is carried out after assigning (I) a vector of initial values to the states of the nodes and (II) a function that transforms the product of the connection matrix with the vector of the initial values into a vector representing the values of the nodes at an instant later. The transformation function can be discrete (such as the

Heaviside function) or continuous (such as the logistic function). In the case of discrete functions, the complex systems can evolve into an attractor constituted by a stable node or limit cycle. In the case of continuous functions, even strange attractor can emerge [17].

Both fuzzy logic systems and fuzzy cognitive maps can be built either by human experts or automatically through learning algorithms. It may happen that the membership functions of the fuzzy sets are not certain but have definite degrees of uncertainty. For these cases, Zadeh introduced [18] the concept of type-2 fuzzy sets that is an extension of the concept of an ordinary fuzzy set, i.e., a type-1 fuzzy set. Type-2 fuzzy sets have grades of membership that are themselves fuzzy. At each value of the primary variable x, the membership is a function and not just a point value: it is the secondary membership function (w). The domain of w is in the interval [0, 1] and its range is also in [0, 1] (see Figure 4). Therefore, the membership function of a type-2 fuzzy set is three-dimensional [19]. If projected on a plane, it gives rise to the footprint of uncertainty, which is bound by a lower membership function (LB) and an upper membership function (UB). In Figure 4, LB and UB are represented as continuous black lines. The footprint of uncertainty embeds the type-1 fuzzy set delimited by dashed lines. Type-2 fuzzy sets find many applications in intelligent control, pattern recognition, intelligent manufacturing, time series prediction, and other fields [20].

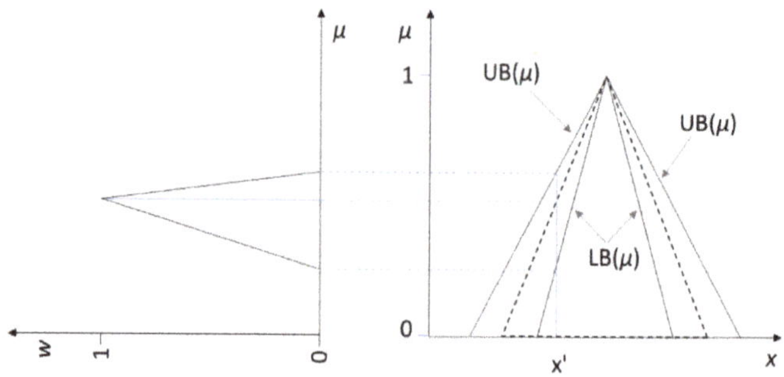

Figure 4. An example of type-2 fuzzy set. The original type-1 fuzzy set is the dashed triangular set. The lower (LB) and upper (UB) bounds define the footprint of uncertainty. The plot on the left shows the trend of the secondary membership (w) when $x = x\prime$

.

3. Fuzzy Logic and the Human Nervous System

Fuzzy logic is a valid model of the human capability to compute with words because there are structural and functional analogies between the human nervous system (HNS) and a Fuzzy logic system [21,22]. The HNS is a complex network of billions of nerve cells distributed throughout our organism [23]. It monitors the environment and our body, and it masters our behavior after collecting information, processing it, taking decisions. The HNS comprises three elements: (I) the sensory system; (II) the central nervous system; (III) the effectors' system. The sensory system catches physical and chemical signals and transduces them in electrochemical information that is sent to the brain. Into the brain, information is integrated, stored and processed. The outputs of the cerebral computations are electro-chemical commands sent to the components of the effectors' system, i.e., glands and muscles. Our sensory system encompasses eight sensory subsystems: a visual system to detect light; an olfactory and a gustatory system to probe chemicals in the air we breathe and in what we uptake through our mouth, respectively; an auditory, tactile, and proprioceptive system provided with mechanoreceptors that perceive either steady or vibrating or instantaneous mechanical forces; thermoreceptors to distinguish cold from warm stimuli; nociceptors to alert our body in the presence of noxious situations. Each sensory subsystem has a hierarchical structure. At the lowest level, there is a collection of receptor proteins. At an upper level, there are receptor cells that contain several replicas of the receptor molecules. We have many copies of the receptor cells properly distributed in space, often covering a tissue. The tissue may be located in an organ provided with an accessory structure that conveys the stimuli to the receptor cells. Every sensory subsystem encodes four aspects of a stimulus: its modality (M), intensity (I_M), spatial distribution ($I_M(x, y, z)$), and time evolution ($I_M(t)$). This multiple information is encoded hierarchically. In fact, the modality is encoded at the molecular level. The ensemble of the molecular receptors of a specific sensory subsystem works as a collection of molecular fuzzy sets: they granulate the modality of the kind of stimulus they sense. Signals that are perceived by the same sensory subsystem but have distinct modalities belong to the collection of the molecular fuzzy sets at different degrees. In other words, the modality of the signals is encoded as fuzzy information at the molecular level through the molecular Fuzzy sets that work in parallel.

An example is shown in Figure 5. It regards our visual system. The modality is the spectral composition of the light. We have three types of photoreceptor proteins, labeled as "Blue", "Green", and "Red", respectively. They allow us to distinguish colors. Their absorption spectra granulate the visible spectral region in three molecular fuzzy sets. Each band is due to the vibrational energies of the lowest excited π^* state of the retinal chromophore. Light beams having distinct spectral compositions belong to the three molecular fuzzy sets at different degrees (in Figure 5, the memberships of a green and a red light are depicted).

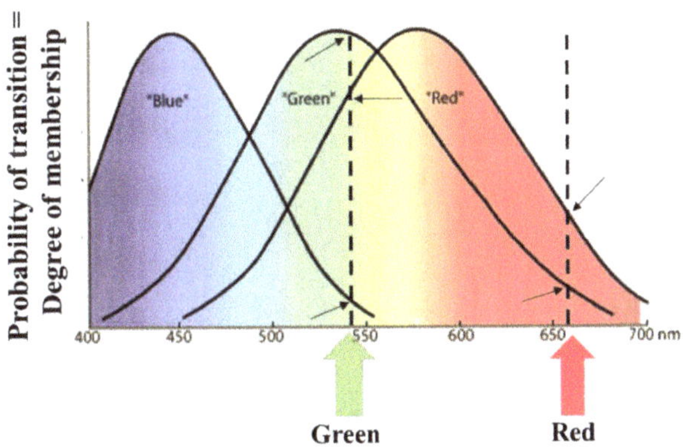

Figure 5. Absorption spectra of the "Blue", "Green", and "Red" photoreceptors that partition the visible spectral region in three fuzzy sets. Beams having different colors belong to the three molecular Fuzzy sets at different degrees. The degrees of membership of one pure green and one pure red beam to three Fuzzy sets are shown (see the arrows).

In living cells, when a stimulus actively interplays with a molecular receptor that is a protein, it promotes its structural change. Within cells, there are several copies of the molecular sensors (see Figure 6A). The number of molecular receptors that are activated in a cell depends on the intensity of the stimulus. Each cell plays like a cellular fuzzy set, and the degree of membership of a stimulus to a cellular fuzzy set encodes the intensity of the stimulus. The molecular structural modifications induced by the stimulus trigger intracellular cascade reactions, finally modifying the electrochemical permeability of the receptor cells membranes. The extent of the change in the electrochemical permeability depends on how many molecular receptors have changed their structure and hence on the intensity of the stimuli. The receptor cells produce graded potentials that are analog signals. The information of such signals is usually converted in the firing rate of the action potential trains. Often, the action potentials are produced by an architecture of afferent neurons that integrate the information regarding the spatial distribution of the stimuli (see Figure 6B). In fact, every afferent neuron has a receptive field that works as a fuzzy set encompassing specific receptor cells. For instance, in the visual subsystem, the photoreceptor cells are granulated by the bipolar cells. Light shining on the center of a bipolar cell's receptive field and light shining on its surround produce opposite changes in the cell's membrane potential. The purpose of the bipolar fuzzy sets is to improve the contrast and definition of the visual stimuli. The center-surround structure of the receptive fields of the bipolar cells is transmitted to the ganglion cells. The accentuation of contrasts by the center-surround receptive fields of the bipolar cells is thereby preserved and passed on to the ganglion cells. The presence of overlapping receptive fields (like overlapping fuzzy sets) allows processing the information of a light stimulus in parallel and increasing the acuity by highlighting the contrasts in space and time. The action potentials generated by the afferent neurons are the ideal code for sending the information up to the brain. In the cerebral cortex, there are areas having different intrinsic rhythms [24–26]. They form a neural dynamic space partitioned in overlapped cortical fuzzy compartments (see Figure 6C). Such cortical fuzzy sets are activated at different degrees by separate attributes of the perceptions and produce a meaningful experience of the external and internal worlds.

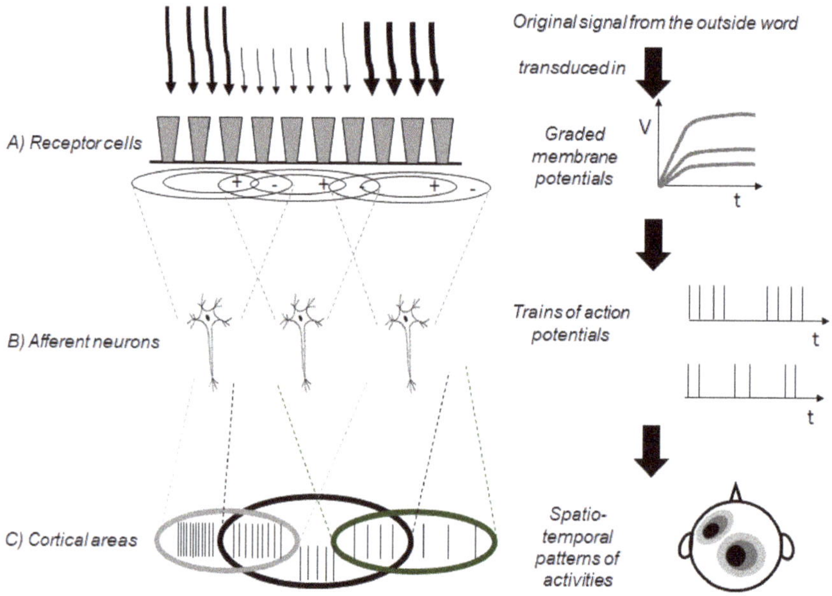

Figure 6. Scheme of the action of a sensory subsystem made of three principal elements described as three collections of fuzzy sets. First, the sensory cellular Fuzzy sets (**A**) that encode the information of a signal as graded potentials. Second, the afferent neurons (**B**) whose receptive fields are fuzzy sets: they encode the information as firing rates of the action potential trains. Third, the cortical areas (**C**) that are partitioned in different dynamic regimes giving rise to an infrastructure of fuzzy sets encoding distinct syntactic and semantic attributes of the original signals.

Based on this description, it might seem that sensory perception is objective, universal, reproducible, and deterministic. However, this is not the case. In fact, sensory perception depends on the physiological state of the perceiver, his/her past experiences, and each sensory system is unique and not universal. Moreover, every human brain must deal with the uncertainty in the perception. Under uncertainty, an efficient way of performing tasks is to represent knowledge with probability distributions and acquire new knowledge by following the rules of the probabilistic inference [27,28]. Therefore, it is reasonable to assume that the human brain performs probabilistic reasoning, and the human perception can be described as a subjective process of Bayesian probabilistic inference [29,30]. In fact, the frequentist probability can be used only in the case of a large number of trials. According to the Bayesian probabilistic inference, the perception of a signal $I_M(x, y, z, t)$ by cortical cells CC_M is given by the "posterior probability" $p(I_M|CC_M)$:

$$p(I_M|CC_M) = \frac{p(CC_M|I_M)p(I_M)}{p(CC_M)}, \tag{1}$$

In (1), $p(CC_M|I_M)$ is the "likelihood", $p(I_M)$ is the "prior probability", and $p(CC_M)$ is the "plausibility". The plausibility is only a normalization factor. In agreement with the theory of Bayesian probabilistic inference generalized in fuzzy context [31], the likelihood may be identified with the hierarchical and deterministic fuzzy information described previously in this paragraph (see also Figure 7). The prior probability $p(I_M)$ comes from the knowledge of the regularities of the signals and represents the influence of the brain on human perception. In fact, human perception is a trade-off between the likelihood and the prior probability [32]. If the likelihood represents the deterministic and objective part of the human perception, on the other hand, the prior probability represents its subjective contribution. The noisier and ambiguous are the features of a signal, the more prior probability driven will be the perception, and the less reproducible and universal will be the sensation.

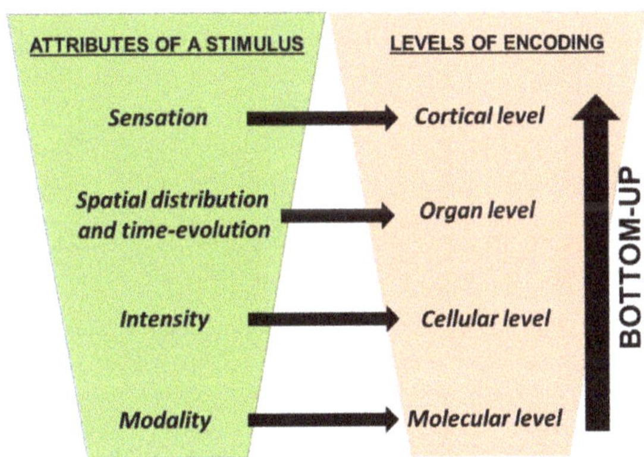

Figure 7. Hierarchical mechanism of encoding the information of a stimulus.

Sometimes, we receive multimodal signals that interact with more than one sensory subsystem. Each activated sensory subsystem produces its own mono-sensory fuzzy information. Physiological and behavioral experiments have shown that the brain integrates the mono-sensory perceptions to generate the final sensation [33]. Multisensory processing pieces signals of different modality if stimuli fall on the same or adjacent receptive fields (according to the "spatial rule") and within close temporal proximity (according to the "temporal rule"). Since sensory modalities are not equally reliable, and their reliability can change with context, multisensory integration involves statistical issues, and it is often assumed to be a Bayesian probabilistic inference [34]. Clearly, the experience of the world is influenced by the past perceptive events, stored in the memory presumably under the shape of fuzzy rules. These stored events and rules confer to the humans the remarkable power of making decisions in complex situations and recognizing variable patterns.

4. The Methodologies to Implement Fuzzy Sets and Process Fuzzy Logic at the Molecular Level

Fuzzy logic is routinely implemented in digital electronic circuits. However, the best accomplishments of FLSs have been achieved through analog electronic circuits. Whereas the digital circuits are based on electrical signals that vary steeply in sigmoid manner, the analog circuits are based on signals that vary smoothly in hyperbolic or linear manner. The analog circuits guarantee the best implementations of an infinite-valued logic that is fuzzy logic.

In the recent years, fuzzy logic has been implemented by using even molecules and chemical reactions. Three principal strategies can be outlined:

The first strategy is an imitation of the sensory subsystems described in the previous paragraph. In every sensory subsystem, there is a collection of distinct sensory cells that works as an ensemble of cellular fuzzy sets embedding molecular fuzzy sets. The cellular fuzzy sets work in parallel. The information of a stimulus is encoded as a vector of degrees of membership of the stimulus to the cellular fuzzy sets. This strategy will be called the "fuzzy parallelism" approach.

The second strategy is an imitation of how the proteins work in the immune and the biomolecular information systems. Almost every protein is a fuzzy set because it exists as an ensemble of many conformers that have context-dependent dynamic behavior. The macromolecular conformers are adaptable and subjected to the laws of the natural selection. They are the "words" of the cellular language. The imitation of the proteins of the cells and the immune system allows to implement the so-called "conformational fuzziness" strategy.

Finally, the third strategy derives from the fuzziness of the quantum world and it will be called "quantum fuzziness". When superimposed quantum states undergo decoherent phenomena, it is possible to exploit heaps of molecules to process fuzzy logic through macroscopic, smooth, analog input and output variables.

Examples of the three strategies are described in the following three subparagraphs.

4.1. The "Fuzzy Parallelism" Approach

In Section 3, we have discovered that the absorption bands of the three photoreceptor proteins present on the fovea of the retina play as three molecular fuzzy sets. Lights that differ in their spectral compositions belong to the three bands at distinct degrees, and they are perceived as different colors. Moreover, the millions of replicas of the three photoreceptor proteins within each photoreceptor cell allow determining the intensity of the signals at every wavelength. The imitation of the way we distinguish colors has allowed the design and implementation of chemical systems that extend human vision to the UV [35,36]. Such chemical systems are based on direct thermally reversible photochromic compounds. A thermally reversible photochromic compound is a species that in the absence of any radiation, it exists in a structure (i.e., A in Figure 8) that absorbs just in the UV and it is uncolored. Upon UV, it transforms in B that also absorbs in the visible region. When B is formed, the system becomes colored (see Figure 8). The transformation of A into B is thermally reversible. In other words, if we discontinue the UV irradiation, the color bleaches because the B molecules transform back to the original structure A, spontaneously at room temperature. Mixtures of properly chosen direct thermally reversible photochromic compounds extend the human capability of distinguishing electromagnetic spectra to the UV region. Such mixtures, called biologically inspired photochromic fuzzy logic (BIPFUL) systems, are designed by the following procedure. First, the absorption bands of the uncolored forms, A_i, are assumed to be input fuzzy sets. Second, the absorption bands of the colored forms, B_i, are assumed to be output fuzzy sets. Third, the algorithm expressing the degree of membership of the UV radiation, having intensity $I_0(\lambda_{irr})$ at the wavelength λ_{irr}, to the absorption band of the A_i compound is:

$$\mu_{UV,A_i} = \Phi_{PC,A_i}(\lambda_{irr}) I_0(_{irr}) \left(1 - 10^{-\varepsilon_{A_i} C_{0,i} l}\right), \tag{2}$$

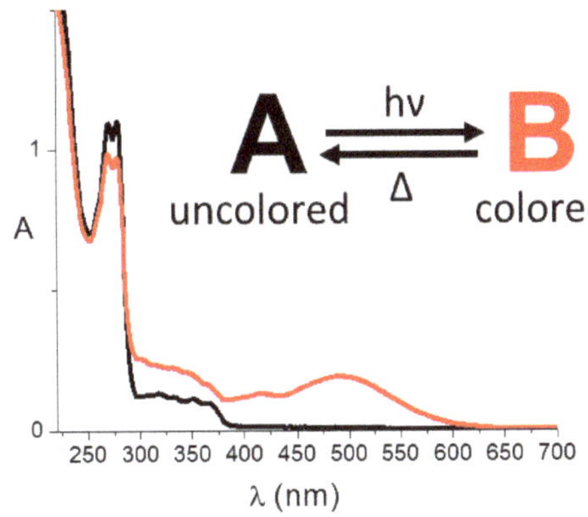

Figure 8. Example of a direct thermally reversible photochromic compound.

In Equation (2), $\Phi_{PC,A_i}(\lambda_{irr})$ is the photochemical quantum yield of photo-coloration for A_i, ε_{A_i} is the absorption coefficient at $_{irr}$ for the A_i photochromic species, and $C_{0,i}$ is its analytical concentration. Finally, the equation expressing the activation of the B_i output fuzzy sets is:

$$A_{B_i} = \frac{\varepsilon_{B_i}(\lambda_{an})}{k_{\Delta,i}} \mu_{UV,A_i}. \quad (3)$$

In Equation (3), A_{B_i} is the absorbance at the wavelength λ_{an} into the visible and due to the coloured form of the i-th photochromic species; $\varepsilon_{B_i}(\lambda_{an})$ is its absorption coefficient, and $k_{\Delta,i}$ is the kinetic constant of the bleaching reaction for B_i. Each absorption spectrum recorded at the photo-stationary state will be the sum of as many terms represented by equation (3) as there are photochromic components within the BIPFUL system. The BIPFUL systems that have been devised are made of naphthopyrans and spiroxazines, and they allow to discriminate the three regions of the UV spectrum, i.e., UV-A, UV-A, UV-B, and UV-C.

The imitation of all the other sensory subsystems, conceived as hierarchical fuzzy systems where a collection of distinct molecular and cellular fuzzy sets work in parallel (see Section 3), should allow to devise artificial sensory systems that have the power of extracting the essential features of stimuli and recognizing variable patterns.

4.2. *"Conformational Fuzziness"*

Within every living cell, there are many proteins that work as if they were the neurons of the "cellular nervous system". They participate in the signaling and genetic networks and allow the cell to respond to the ever-changing environmental conditions. Specific proteins, called antibodies, are also the fundamental ingredients of the immune system that protects our bodies from intruders. A limited set of flexible antibodies can bind a wide range of antigens. Proteins are ubiquitous in living beings and they play multiple roles, due to their "dynamism and evolvability" [37]. In fact, proteins are conformationally dynamic and exhibit functional promiscuity. Conformational dynamism and heterogeneity enable context-specific functions to emerge in response to changing environmental conditions and, furthermore, allow a single structural motif to be used in multiple settings [38]. The conformational flexibility and heterogeneity of proteins represent their fuzziness.

Conformational fuzziness is not a prerogative feature of proteins. Even the long polymer of chromatin in the nucleus of eukaryotic cells is Fuzzy. Some portions contain heterochromatin made of DNA packed tightly around histones. Some other areas contain euchromatin that is DNA loosely packed. Usually, genes in euchromatin are active, whereas those in heterochromatin are inactive. Euchromatin exposes a broader and rougher surface to the proteins scanning for their target sequences.

Heterochromatin is flatter, smoother, and with a less extended surface [39]. Chromatin organization is highly dynamic, varying both during the cell cycle and among different cell types [40].

Conformational fuzziness is not unique to macromolecules, but it can be experienced even with simple molecules. An example is the fuzziness of the merocyanine (MC) that is generated by UV irradiation of the spirooxazine (SpO) shown in Figure 9 [41]. Since MC has a flexible molecular skeleton, it gives rise to many conformers. The number and type of conformers depend on the physical and chemical context (for example, temperature, solvent, and the presence of a docking glycine).

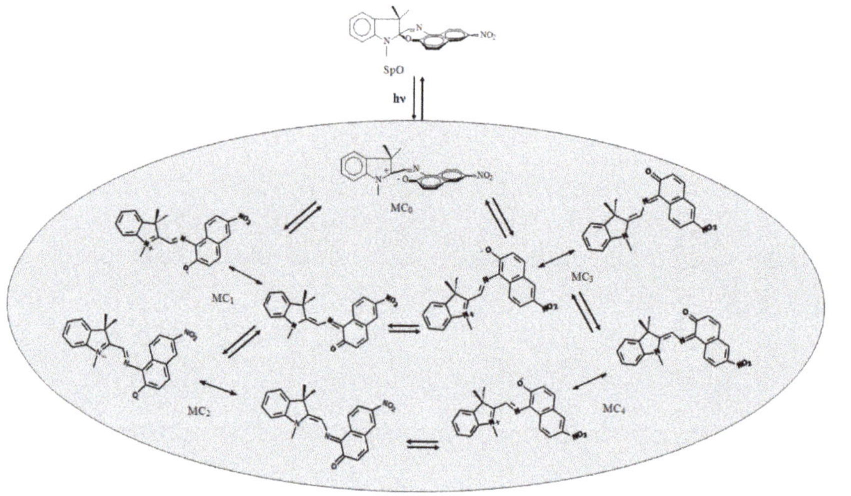

Figure 9. Just of a few of all the possible conformers of a merocyanine (MC_i) produced by irradiation of a spirooxazine (SpO).

Whatever the compound is, being either a macromolecule or a molecule, the ensemble of its conformers plays like a molecular Fuzzy set. Its fuzziness may be quantified by determining its fuzzy entropy. A definition of fuzzy entropy based on Shannon's function of information entropy is [42,43]:

$$H = -K \sum_{i=1}^{n} (\mu_i \log_{10}(\mu_i) + (1 - \mu_i) \log_{10}(1 - \mu_i)), \quad (4)$$

where μ_i is the relative weight of the i-th conformer, n is the total number of conformers, and $K = (1/n)$ is a normalization factor. The fuzzy entropy of a compound is context-dependent, like the meaning of a word in natural language. In fact, conformationally heterogeneous structures are adaptable to many different contexts. Of course, the fuzzy entropy of a macromolecule is significantly larger than that of a simple molecule. Among proteins, those completely or partially disordered [44] are the fuzziest. Their pronounced fuzziness makes them multifunctional and even able to moonlight [45], i.e., play distinct functions, depending on their context.

4.3. *"Quantum Fuzziness"*

Isolated microscopic systems exist in a superposition of states. For instance, if there are two accessible states, indicated as $|0\rangle$ and $|1\rangle$, the isolated microscopic system exists in a quantum state $|\Psi\rangle$ that is a linear combination of $|0\rangle$ and $|1\rangle$:

$$|\Psi\rangle = a|0\rangle + b|1\rangle, \quad (5)$$

where a and b are complex numbers that verify the normalization condition $|a|^2 + |b|^2 = 1$. The states $|0\rangle$ and $|1\rangle$ can be imagined as two fuzzy sets. Their vagueness, i.e., their fuzziness is outlined by the Heisenberg's Uncertainty Principle. The $|\Psi\rangle$ state belongs to $|0\rangle$ and $|1\rangle$ with degrees that are $|a|^2$ and $|b|^2$, respectively. $|\Psi\rangle$ is a qubit, i.e., the elementary unit of the quantum information. The qubit can be described as a unit vector in a two-dimensional Hilbert space. The state of the qubit can be also represented by the following equation:

$$|\Psi\rangle = \cos\left(\frac{\theta}{2}\right)|0\rangle + e^{i\varphi} sin\left(\frac{\theta}{2}\right)|1\rangle \tag{6}$$

where θ and φ define a point on the unit three-dimensional sphere, called the Bloch sphere. Logic operations on qubits can be visualized as rotations of the unit vectors on the Bloch sphere, preserving the norm of the quantum states. If a microscopic system is a superposition of n qubits, it has 2^n accessible states, simultaneously. If we make an operation on this system, we manipulate 2^n states, at the same time. Therefore, it is evident the alluring computational power of quantum logic. However, the main difficulty is to avoid the decoherence of the superimposed quantum states, which can be induced by deleterious interactions with the surrounding environment [46]. The decoherence induces the collapse of a qubit in one of its two originally accessible states, either $|0\rangle$ or $|1\rangle$, with probabilities $|a|^2$ and $|b|^2$, respectively. Whenever the decoherence is unavoidable, the single particles can be used to process discrete logics, i.e., binary or multi-valued crisp logics [47,48]. Of course, specific microscopic techniques, reaching the atomic resolution, are needed to carry out the computations. Alternatively, we may think of making computations by exploiting large assemblies of particles, e.g., molecules. Vast collections of molecules (amounting to the order of the Avogadro's number) appear as bulky materials. The inputs and outputs for making computations become macroscopic variables that can change in a continuous manner. The relations establishing between the inputs and the outputs can be either steep or smooth. Steep, sigmoid functions are suitable to implement discrete logics, whereas both linear and nonlinear smooth functions are suitable to build fuzzy logic systems [49]. Some fuzzy logic gates and operations have been implemented by the hybridization reaction of DNA [50,51] and the supramolecular interactions between carbohydrates and proteins [52]. Other fuzzy logic systems have been built by exploiting the dependence of the fluorescence quantum yield on physical and chemical inputs. One example is the dependence of the fluorescence of 6(5*H*)-phenanthridinone (see Figure 10A) on the hydrogen bonding donation ability of the solvent (HBD) and the temperature [53]. Another example is given by tryptophan, both as isolated molecule and bonded to the serum albumin, whose fluorescence depends on the temperature and the amount of the quencher flindersine (see Figure 10B) [54]. A further example is a ruthenium complex, whose fluorescence depends on Fe^{2+} and F^- (see Figure 10C) [55]. A final example is the fluorescence of europium bound to a metal-organic framework, which depends on metal cations, such as Hg^{2+} and Ag^+ (see Figure 10D) [56]. The emission of light is a preferable output because it bridges the gap between the microscopic and the macroscopic world. A multi-responsive chromogenic compound, belonging to the class of spirooxazine, has been used for the implementation of the all fundamental fuzzy logic gates, AND, OR, and NOT [57]. The protons, Cu^{2+}, and Al^{3+} ions were used as inputs, and the color coordinates (R, G, B) or the colorability [41] of the chromogenic compound as outputs. Then, other platforms have been proposed. For example, a multi-state tantalum oxide memristive device [58] and an anthraquinone-modified titanium dioxide electrode [59]. Even, the Belousov-Zhabotinsky reaction, carried out in oscillatory regime and in an open system [60], allows to implement all the fundamental fuzzy logic gates by using bromide and silver ions as chemical inputs and the period of the oscillations as outputs. Finally, the "hydrodynamic photochemical oscillator", which is a thermally reversible photochromic compound combined with the convective motion of the solvent, is suitable to implement fuzzy logic systems when it works in chaotic regime [61]. All these examples show that fuzzy logic can be processed not only by conventional electronic circuits but also by unconventional chemical systems exhibiting analog input-output relationships in either the liquid or the solid phase.

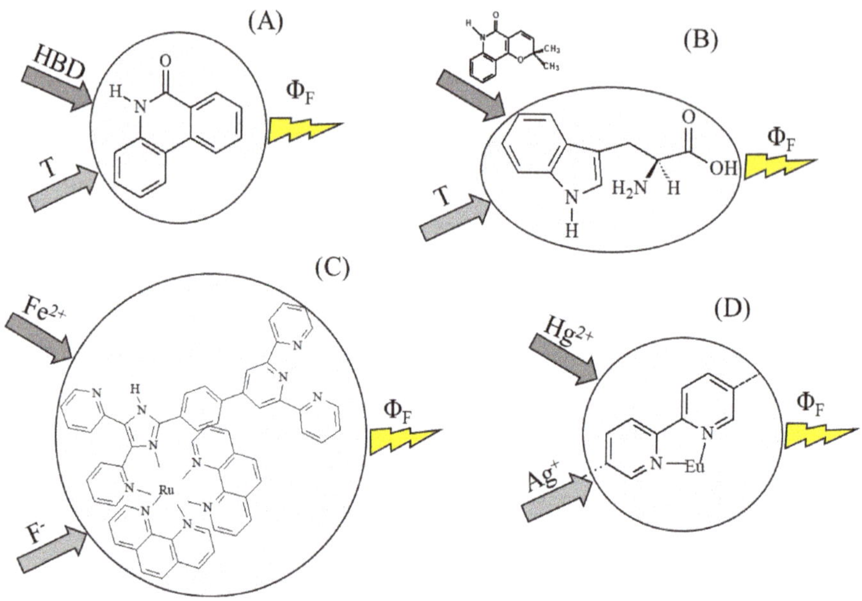

Figure 10. Dependence of the fluorescence quantum yield of 6(5*H*)-phenanthridinone (**A**), tryptophan (**B**), a ruthenium complex (**C**), and europium bounded to a metal-organic framework (**D**) on physical and chemical inputs.

5. Perspectives of the Fuzziness of the Molecular World

Fuzzy logic is a valid model of the human power to compute with words and take decisions in complex situations. The closer one looks at the real-world problems, the fuzzier become their solutions. Fuzzy logic is playing a relevant role in the field of artificial intelligence when we deal with complex systems.

This work highlights that even the molecular world is fuzzy. In fact, quantum logic is fuzzy ("quantum fuzziness"). A qubit is a superposition of two distinct quantum states that are like fuzzy sets. Therefore, quantum logic might be considered a particular kind of fuzzy logic. When decoherent phenomena induce the collapse of qubits, it is not possible to process quantum logic. However, by working with large collections of molecules, it is feasible to implement fuzzy logic systems, when causal, macroscopic, smooth, analog input-output relationships are found.

In the human sensory system, the sensory cells that are fuzzy sets, containing molecular fuzzy sets, collect a large amount of data. The hierarchical architecture of the afferent and cortical neurons, which is based on the overlapping of their receptive fields, allows extracting only the meaningful information of the big data contained in the stimuli. The imitation of the principal features of the sensory system, in particular of what we called as "fuzzy parallelism", should allow devising artificial sensory system able to extract the essential characteristics of the complex stimuli. Hence, such artificial sensory systems should be suitable to recognize variable patterns.

The computational power of the cells and the human immune system derives from the "conformational fuzziness" of their macromolecules. By exploiting the conformational elasticity of molecules, especially proteins, it is possible to process fuzzy logic. In fact, the "conformational fuzziness" makes molecules adaptable to their microenvironment. This feature is suitable to implement the dependence of the information on the context.

By processing fuzzy logic at the molecular level, we want to promote the development of the chemical artificial intelligence. The purpose of chemical artificial intelligence is to mimic the performances of the human intelligence by using not software or hardware, but rather chemical and photochemical reactions in wetware. In fact, there exist chemical systems that can work as surrogates of the neural dynamics [62–65]. These systems can interact and communicate by exploiting both chemical and electrical and optical signals. They are the fundamental components of a futuristic opto-/electro-brain-like computing machine that should be suitable to recognize variable patterns and compute with words. There is a long path before the concrete implementation of this new generation of computing machines, more similar to the brain rather than to the electronic computer from both the structural and the functional point of view. Further analysis of the human nervous system and further development of the theory of fuzzy logic are needed. For example, the receptive field of a neuron can inspire a new kind of fuzzy set (i.e., Type-III fuzzy set) where we distinguish inhibitory and excitatory actions. With this new kind of fuzzy set, implemented somehow artificially, the recognition of variable patterns should become easier. Moreover, the chemical artificial intelligence will boost the development of the soft robotics. Soft robots, also called "chemical robots", will be easily miniaturized and implanted in living beings [66–71]. They will interplay with cells and organelles for biomedical applications. They will become auxiliary elements of the human immune system to defeat diseases that are still incurable.

Finally, this field of research could give clues about the origin of the life on Earth. In fact, the appearance of the life on Earth, occurred roughly 3.5 billion of years ago, was like a "phase transition". It was a transition from inanimate chemical systems, unable to encode, process, communicate and store information, to the living chemical systems, able to exploit the matter and energy to encode, process, send, and store information. The development of chemical artificial intelligence could unveil how that unique "phase transition" happened.

Funding: Please add: "This research was funded by ANVUR grant number [n.20/2017]."

Acknowledgments: P.L. Gentili is grateful to his stricter collaborators, who are B.M. Heron of the Huddersfield University (UK), R. Germani of the University of Perugia (Italy), I.R. Epstein and M. Dolnik of the Brandeis University (MA, USA), H. Gotoda of the Tokyo University of Science (Japan), and J.-C. Micheau of the Université Paul Sabatier-Toulouse III (France).

Conflicts of Interest: The author declares no conflict of interest.

References

1. Marx, V. The Big Challenges of Big Data. *Nature* **2013**, *498*, 255–260. [CrossRef] [PubMed]
2. Whitfield, J.D.; Love, P.J.; Aspuru-Guzik, A. Computational complexity in electronic structure. *Phys. Chem. Chem. Phys.* **2013**, *15*, 397–411. [CrossRef] [PubMed]
3. Bishop, C.M. *Pattern Recognition and Machine Learning*; Springer: New York, NY, USA, 2006.
4. Gentili, P.L. *Untangling Complex Systems: A Grand Challenge for Scienc*; CRC Press, Taylor & Francis Group: Boca Raton, FL, USA, 2018.
5. Gentili, P.L. A strategy to face complexity: The development of chemical artificial intelligence. In *Advances in Artificial Life, Evolutionary Computation, and Systems Chemistry*; Rossi, F., Piotto, S., Concilio, S., Eds.; Springer: Cham, Switzerland; New York, NY, USA, 2017; Volume 708, pp. 151–160.
6. Albus, J.S.; Bekey, G.A.; Holland, J.H.; Kanwisher, N.G.; Krichmar, J.L.; Mishkin, M.; Modha, D.S.; Raichle, M.E.; Shepherd, G.M.; Tononi, G. A proposal for a Decade of the Mind initiative. *Science* **2007**, *317*, 1321. [CrossRef] [PubMed]
7. Zadeh, L.A. Outline of a New Approach to the Analysis of Complex Systems and Decision Processes. *IEEE Trans. Syst. Man Cyb.* **1973**, *3*, 28–44. [CrossRef]
8. Zadeh, L.A. A New Direction in AI. Toward a Computational Theory of Perceptions. *AI Mag.* **2001**, *22*, 73–84.
9. Zadeh, L.A. Toward Human Level Machine Intelligence-Is It Achievable? The Need for a Paradigm Shift. *IEEE Comput. Intell. Mag.* **2008**, *3*, 11–22. [CrossRef]
10. Gentili, P.L. Small steps towards the development of chemical artificial intelligent systems. *RSC Adv.* **2013**, *3*, 25523–25549. [CrossRef]
11. Zadeh, L.A. Fuzzy sets. *Inform. Control* **1965**, *8*, 338–353. [CrossRef]

12. Mendel, J.M. Fuzzy logic systems for engineering: A tutorial. *Proc. IEEE* **1995**, *83*, 345–377. [CrossRef]
13. Li, Z.; Chen, G. *Integration of Fuzzy Logic and Chaos Theory*; Springer Science & Business Media: Dordrecht, The Netherlands, 2006.
14. Gentili, P.L.; Gotoda, H.; Dolnik, M.; Epstein, I.R. Analysis and prediction of aperiodic hydrodynamic oscillatory time series by feed-forward neural networks, fuzzy logic, and a local nonlinear predictor. *Chaos* **2015**, *25*, 013104. [CrossRef] [PubMed]
15. Kosko, B. Fuzzy cognitive maps. *Int. J. Man-Mach. Stud.* **1986**, *24*, 65–75. [CrossRef]
16. Axelrod, R. *Structure of Decision: The Cognitive Maps of Political Elites*; Princeton University Press: Princeton, NJ, USA, 1976.
17. Stach, W.; Kurgan, L.; Pedrycz, W.; Reformat, M. Genetic learning of fuzzy cognitive maps. *Fuzzy Sets Syst.* **2005**, *153*, 371–401. [CrossRef]
18. Zadeh, L.A. The concept of a linguistic variable and its application to approximate reasoning-1. *Inf. Sci.* **1975**, *8*, 199–249. [CrossRef]
19. Mendel, J.M. Type-2 Fuzzy Sets and Systems: An Overview. *IEEE Comput. Intell. Mag.* **2007**, *2*, 20–29. [CrossRef]
20. Castillo, O.; Melin, P. *Type-II Fuzzy Logic: Theory and Applications*; Springer-Verlag: Heidelberg, Germany, 2008.
21. Gentili, P.L. The human sensory system as a collection of specialized fuzzifiers: A conceptual framework to inspire new artificial intelligent systems computing with words. *J. Intell. Fuzzy Syst.* **2014**, *27*, 2137–2151.
22. Sandler, U.; Tsitolovsky, L. *Neural Cell Behavior and Fuzzy Logic*; Springer: New York, NY, USA, 2008.
23. Paxinos, G.; Mai, J.K. *The Human Nervous System*, 2nd ed.; Elsevier: San Diego, CA, USA, 2004.
24. Johnson, E.; Hawken, M.J.; Shapley, R. The spatial transformation of color in the primary visual cortex of the macaque monkey. *Nat. Neurosci.* **2001**, *4*, 409–416. [CrossRef] [PubMed]
25. Friedmann, H.S.; Zhou, H.; von der Heydt, R. The coding of uniform color figures in monkey visual cortex. *J. Physiol. (London)* **2003**, *548*, 593–613. [CrossRef] [PubMed]
26. Gegenfurtner, K.R. Cortical Mechanisms of Colour Vision. *Nat. Rev. Neurosci.* **2003**, *4*, 563–572. [CrossRef] [PubMed]
27. Van Horn, K.S. Constructing a logic of plausible inference: A guide to Cox's theorem. *Int. J. Approx. Reason.* **2003**, *34*, 3–24. [CrossRef]
28. De Finetti, B.; Machi, A.; Smith, A. *Theory of Probability: A Critical Introductory Treatment*; John Wiley & Sons: New York, NY, USA, 1992.
29. Ma, W.J.; Beck, J.M.; Pouget, A. Spiking networks for Bayesian inference and choice. *Curr. Opin. Neurobiol.* **2008**, *18*, 217–222. [CrossRef] [PubMed]
30. Mach, E. *Contributions to the Analysis of the Sensations*; Open Court Publishing Co.: Chicago, IL, UAS, 1984.
31. Coletti, G.; Scozzafava, R. Conditional probability, fuzzy sets, and possibility: A unifying view. *Fuzzy Sets Syst.* **2004**, *144*, 1–227. [CrossRef]
32. Kersten, D.; Mamassian, P.; Yuille, A. Object Perception as Bayesian Inference. *Annu. Rev. Psychol.* **2004**, *55*, 271–304. [CrossRef] [PubMed]
33. Ernst, M.O.; Banks, M.S. Humans integrate visual and haptic information in a statistically optimal fashion. *Nature* **2002**, *415*, 429–433. [CrossRef] [PubMed]
34. Pouget, A.; Deneve, S.; Duhamel, J.-R. A computational perspective on the neural basis of multisensory spatial representations. *Nat. Rev. Neurosci.* **2002**, *3*, 741–747. [CrossRef] [PubMed]
35. Gentili, P.L.; Rightler, A.L.; Heron, B.M.; Gabbutt, C.D. Extending human perception of electromagnetic radiation to the UV region through biologically inspired photochromic fuzzy logic (BIPFUL) systems. *Chem. Commun.* **2016**, *52*, 1474–1477. [CrossRef] [PubMed]
36. Gentili, P.L.; Rightler, A.L.; Heron, B.M.; Gabbutt, C.D. Discriminating between the UV-A., UV-B and UV-C regions by novel Biologically Inspired Photochromic Fuzzy Logic (BIPFUL) systems: A detailed comparative study. *Dyes Pigm.* **2016**, *135*, 169–176. [CrossRef]
37. Tokuriki, N.; Tawfik, D.S. Protein Dynamism and Evolvability. *Science* **2009**, *324*, 203–207. [CrossRef] [PubMed]
38. Pricer, R.; Gestwicki, J.E.; Mapp, A.K. From Fuzzy to Function: The New Frontier of Protein-Protein Interactions. *Acc. Chem. Res.* **2017**, *50*, 584–589. [CrossRef] [PubMed]

39. Bancaud, A.; Huet, S.; Daigle, N.; Mozziconacci, J.; Beaudouin, J.; Ellenberg, J. Molecular crowding affects diffusion and binding of nuclear proteins in heterochromatin and reveals the fractal organization of chromatin. *EMBO J.* **2009**, *28*, 3785–3798. [CrossRef] [PubMed]
40. Dekker, J.; Rippe, K.; Dekker, M.; Kleckner, N. Capturing Chromosome Conformation. *Science* **2002**, *295*, 1306–1311. [CrossRef] [PubMed]
41. Gentili, P.L. The fuzziness of a chromogenic spirooxazine. *Dyes Pigm.* **2014**, *110*, 235–248. [CrossRef]
42. De Luca, A.; Termini, S. A definition of nonprobabilistic entropy in the setting of fuzzy set theory. *Inf. Control* **1972**, *20*, 301–312. [CrossRef]
43. Al-Sharhan, S.; Karray, F.; Gueaieb, W.; Basir, O. Fuzzy entropy: A brief survey. In Proceedings of the IEEE International Fuzzy Systems Conference, Melbourne, Australia, 2–5 December 2001; pp. 1135–1139.
44. Fuxreiter, M.; Tompa, P. *Fuzziness. Structural Disorder in Protein Complexes*; Springer: Austin, TX, USA, 2012.
45. Jeffery, C.J. An introduction to protein moonlighting. *Biochem. Soc. Trans.* **2014**, *42*, 1679–1683. [CrossRef] [PubMed]
46. Zurek, W.H. Decoherence and the Transition from Quantum to Classical. *Phys. Today* **1991**, *44*, 36–44. [CrossRef]
47. Szaciłowski, K. Digital Information Processing in Molecular Systems. *Chem. Rev.* **2008**, *108*, 3481–3548. [CrossRef] [PubMed]
48. De Silva, A.P. *Molecular Logic-based Computation*; Royal Society of Chemistry: Cambridge, UK, 2013.
49. Gentili, P.L. Molecular Processors: From Qubits to Fuzzy Logic. *ChemPhysChem.* **2011**, *12*, 739–745. [CrossRef] [PubMed]
50. Zadegan, R.M.; Jepsen, M.D.E.; Hildebrandt, L.L.; Birkedal, V.; Kjems, J. Construction of a fuzzy and Boolean Logic Gates Based on DNA. *Small* **2015**, *11*, 1811–1817. [CrossRef] [PubMed]
51. George, A.K.; Singh, H. DNA Implementation of Fuzzy Inference Engine: Towards DNA Decision-Making Systems. *IEEE Trans. Nanobiosci.* **2017**, *16*, 773–782. [CrossRef] [PubMed]
52. Bavireddi, H.; Bharate, P.; Kikkeri, R. Use of Boolean and fuzzy logics in lactose glycocluster research. *Chem. Commun.* **2013**, *49*, 9185–9187. [CrossRef] [PubMed]
53. Gentili, P.L. Boolean and fuzzy logic implemented at the molecular level. *Chem. Phys.* **2007**, *336*, 64–73. [CrossRef]
54. Gentili, P.L. Boolean and Fuzzy Logic Gates Based on the Interaction of Flindersine with Bovine Serum Albumin and Tryptophan. *J. Phys. Chem. A* **2008**, *112*, 11992–11997. [CrossRef] [PubMed]
55. Karmakar, S.; Nandi, M.; Mukherjee, S.; Baitalik, S. Polypyridyl-imidazole based smart Ru(II) complex mimicking advanced Boolean and Fuzzy logic functions. *Inorg. Chim. Acta* **2017**, *454*, 76–88. [CrossRef]
56. Xu, X.-Y.; Yan, B. Intelligent Molecular Searcher from Logic Computing Network Based on Eu(III) Functionalized UMOFs for environmental Monitoring. *Adv. Funct. Mater.* **2017**, *27*, 1700247. [CrossRef]
57. Gentili, P.L. The fundamental Fuzzy logic operators and some complex Boolean logic circuits implemented by the chromogenism of a spirooxazine. *Phys. Chem. Chem. Phys.* **2011**, *13*, 20335–20344. [CrossRef] [PubMed]
58. Bhattacharjee, D.; Kim, W.; Chattopadhyay, A.; Waser, R.; Rane, V. Multi-valued and Fuzzy Logic Realization using TaOx Memristive Devices. *Sci. Rep.* **2018**, 8. [CrossRef] [PubMed]
59. Blachecki, A.; Mech-Piskorz, J.; Gajewska, M.; Mech, K.; Pilaczyk, K.; Szaciłowski, K. Organotitania-Based Nanostructures as a Suitable Platform for the Implementation of Binary, Ternary, and Fuzzy Logic Systems. *ChemPhysChem* **2017**, *18*, 1798–1810. [CrossRef] [PubMed]
60. Gentili, P.L.; Horvath, V.; Vanag, V.K.; Epstein, I.R. Belousov-Zhabotinsky "chemical neuron" as a binary and fuzzy logic processor. *Int. J. Unconv. Comput.* **2012**, *8*, 177–192.
61. Gentili, P.L.; Giubila, M.S.; Heron, B.M. Processing Binary and Fuzzy Logic by Chaotic Time Series Generated by a Hydrodynamic Photochemical Oscillator. *ChemPhysChem* **2017**, *18*, 1831–1841. [CrossRef] [PubMed]
62. Gentili, P.L.; Giubila, M.S.; Germani, R.; Romani, A.; Nicoziani, A.; Spalletti, A.; Heron, B.M. Optical Communication among Oscillatory Reactions and Photo-Excitable Systems: UV and Visible Radiation Can Synchronize Artificial Neuron Models. *Angew. Chem. Int. Ed.* **2017**, *56*, 7535–7540. [CrossRef] [PubMed]
63. Gentili, P.L.; Giubila, M.S.; Germani, R.; Heron, B.M. Photochromic and luminescent compounds as artificial neuron models. *Dyes Pigm.* **2018**, *156*, 149–159. [CrossRef]

64. Zlotnik, A.; Nagao, R.; Kis, I.Z.; Li, S., Jr. Phase-selective entrainment of nonlinear oscillator ensembles. *Nat. Commun.* **2016**, *7*, 10788. [CrossRef] [PubMed]
65. Horvath, V.; Gentili, P.L.; Vanag, V.K.; Epstein, I.R. Pulsed-Coupled Chemical Oscillators with Time Delay. *Angew. Chem. Int. Ed.* **2012**, *51*, 6878–6881. [CrossRef] [PubMed]
66. Hagiya, M.; Konagaya, A.; Kobayashi, S.; Saito, H.; Murata, S. Molecular robots with sensors and intelligence. *Acc. Chem. Res.* **2014**, *47*, 1681–1690. [CrossRef] [PubMed]
67. Adamatzky, A.; de Lacy Costello, B.; Melhuish, C.; Ratcliffe, N. Experimental implementation of mobile robot taxis with onboard Belousov-Zhabotinsky chemical medium. *Mater. Sci. Eng.* **2004**, *24*, 541–548. [CrossRef]
68. Hess, H.; Ross, J.L. Non-equilibrium assembly of microtubules: From molecules to autonomous chemical robots. *Chem. Soc. Rev.* **2017**, *46*, 5570–5587. [CrossRef] [PubMed]
69. Grančič, P.; Štěpánek, F. *Chemical swarm robots in Handbook of Collective Robotics–Fundamentals and Challenges*; Kernbach, S., Ed.; Pan Stanford Publishing: Singapore, 2011.
70. Lund, K.; Manzo, A.J.; Dabby, N.; Michelotti, N.; Johnson-Buck, A.; Nangreave, J.; Taylor, S.; Pei, R.; Stojanovic, M.N.; Walter, N.G.; et al. Molecular robots guided by prescriptive landscapes. *Nature* **2010**, *465*, 206–210. [CrossRef] [PubMed]
71. Kassem, S.; Lee, A.T.L.; Leigh, D.A.; Markevicius, A.; Solà, J. Pick-up, transport and release of a molecular cargo using a small-molecule robotic arm. *Nat. Chem.* **2016**, *8*, 138–143. [CrossRef] [PubMed]

molecules

MDPI

Article

Hardware Realization of the Pattern Recognition with an Artificial Neuromorphic Device Exhibiting a Short-Term Memory

Dawid Przyczyna †, Maria Lis †, Kacper Pilarczyk * and Konrad Szaciłowski *

Academic Centre for Materials and Nanotechnology, AGH University of Science and Technology, al. Mickiewicza 30, 30-059 Kraków, Poland

* Correspondence: kpilarcz@agh.edu.pl (K.P.); szacilow@agh.edu.pl (K.S.)

† These authors contributed equally.

Received: 20 June 2019; Accepted: 26 July 2019; Published: 28 July 2019

Abstract: Materials exhibiting memory or those capable of implementing certain learning schemes are the basic building blocks used in hardware realizations of the neuromorphic computing. One of the common goals within this paradigm assumes the integration of hardware and software solutions, leading to a substantial efficiency enhancement in complex classification tasks. At the same time, the use of unconventional approaches towards signal processing based on information carriers other than electrical carriers seems to be an interesting trend in the design of modern electronics. In this context, the implementation of light-sensitive elements appears particularly attractive. In this work, we combine the abovementioned ideas by using a simple optoelectronic device exhibiting a short-term memory for a rudimentary classification performed on a handwritten digits set extracted from the Modified National Institute of Standards and Technology Database (MNIST)(being one of the standards used for benchmarking of such systems). The input data was encoded into light pulses corresponding to black (ON-state) and white (OFF-state) pixels constituting a digit and used in this form to irradiate a polycrystalline cadmium sulfide electrode. An appropriate selection of time intervals between pulses allows utilization of a complex kinetics of charge trapping/detrapping events, yielding a short-term synaptic-like plasticity which in turn leads to the improvement of data separability. To the best of our knowledge, this contribution presents the simplest hardware realization of a classification system capable of performing neural network tasks without any sophisticated data processing.

Keywords: photoelectrochemistry; wide bandgap semiconductor; artificial neuron; in materio computing; neuromorphic computing

1. Introduction

Pattern recognition is one of the basic cognitive functions, which, due to its complexity and required accuracy, has challenged researchers for decades in a strive to mimic it in an artificial setup. The development of such systems is fueled by various possible applications in medicine, security, economics and many other fields of human's activity. At the moment, the majority of available solutions are based on various software implementations of the machine learning approach, including above all the use of artificial neural networks (ANN) of different architectures. In most of the cases, the ANNs principle of operation is based on the optimization of weights associated with individual connections between nodes (neurons) and the information flow is inspired by the functions of biological structures found in the nervous system. It has been proven on numerous occasions that these algorithms provide an excellent efficiency in various classification tasks with both supervised and unsupervised learning procedures [1–3].

In spite of this, the use of software implementations for ANN algorithms often requires heavy pre- and post-processing of the analysed data and/or a high degree of network complexity translating into a high energy consumption [4]. The use of ANN-based methods may also be associated with a potentially low tolerance towards deliberate attacks [5], emphasized in the case of the one-pixel attack capable of deceiving certain deep neural networks [6]. To meet the discussed drawbacks, some researchers propose the development of hardware implementations for neural networks architectures, incorporating novel materials, non-classic electronic elements and unconventional computing paradigms, such as multi-valued and fuzzy logic systems [7].

The development of neuromorphic computing (NC) can be perceived as one of the manifestations of this trend. The key idea here is to design a brain-inspired hardware computing platform which is optimized towards the implementation of selected aspects of ANN algorithms [7,8]. Among the most advantageous concepts, the use of circuitry employing spiking artificial synapses has been proven to be more energy efficient than the implementation in silico [8]. The construction of these systems is currently being investigated in terms of new materials [9–11] which are applied within integrated networks capable of performing sophisticated information processing [12–14]. At the same time, a number of studies aim at simplifying the circuitry realizing the neuromorphic computations. Wang et al. [15] demonstrated a hybrid convolutional neural network with only one spiking synapse based on a HfO_2 memristor. The system was capable of recognizing handwritten digits with the efficiency of 784 neurons. It was achieved through the time division multiplexing access technique. Nonetheless, the "network" required multilayer information pre-processing and several thousands of software neurons to operate.

The research on photochemical and photoelectrochemical in materio computing devices indicates the possibility of their integration into larger computing systems with the use of optical [16,17] or electrical [18] signals. This can lead to the construction of more complex photoelectrochemical circuits [18], molecular arithmetic-logic units [17] or molecular-scale neural networks [16] and communication systems [19] capable of the sophisticated information processing. The studies on artificial photoelectrochemical synapses [20], devices that may realize elementary learning processes (e.g., paired pulse facilitation, PPF), stimulated further development of the neuromorphic systems combining the neuromimetic approach towards data processing with in materio computing concepts [21–25]. The operation principle of a photoelectrochemical synapse is based on the competition between light-induced charge carrier generation, charge carrier trapping and other interfacial processes affecting the photocurrent generation. Whereas the information processing realized with the use of unconventional molecular or nanoscale devices has several drawbacks compared to classic, silicon-based electronics (usually low speed, some problems with data encoding and concatenation) [26], its combination with classic techniques and algorithms seems to be promising. The term heterotic computation encompasses hybrid systems, in which information processing is performed on various platforms depending on the optimal scenario, utilizing the speed and maturity of in silico computing or the flexibility of the unconventional approaches [27,28].

Here, we present an extremely simplified, robust circuit made of only one photoelectrochemical element, the operation of which is similar to a simple classification system. In the discussed case an Modified National Institute of Standards and Technology Database (MNIST) set of handwritten digits serves as the input data under consideration [29]—without the use of any data pre-processing or software ANNs. The presented optoelectronic device realizes the paired-pulse facilitation (PPF)—a type of short-term plasticity (STP)—seen as an enhancement of the postsynaptic current resulting from the increase in stimulating events frequencies [30,31]. Therefore, we are testing the information processing capability of a single artificial neuron made of nanocrystalline cadmium sulfide. The obtained results show, that the use of a such simple system may improve the separation in the phase space based solely on the characteristics of the input data (unsupervised learning). It seems possible that the proposed approach could be scaled up and a network of similar, interconnected devices could serve as a complex hardware neural network implementing the fuzzy logic formalism and selected concepts of reservoir computing [32].

2. Results and Discussion

2.1. Material Characterization

In order to determine the band gap width (E_g) of cadmium sulfide (CdS), the reflectance spectrum was recorded. Kubelka–Munk's function F_{KM} was calculated based on the raw data and a Tauc plot was made (Figure 1a). CdS is typically considered a direct semiconductor with the E_g value equal to 2.42 eV for the hexagonal phase and 2.33 eV for the tetragonal polymorph. The value determined for the discussed material (2.33 eV, Figure 1a) may suggest the dominance of the latter polymorph, but this value is usually also observed for mixtures of both crystalline phases [33].

The powder X-ray diffraction measurements have been employed to assess the CdS sample composition. The obtained data was analyzed using HighScore Plus software [34] in which so called Rietveld refinement was applied [35]. This method allows evaluation of certain parameters including the volume fraction of phases. The analysis conducted for the diffraction pattern shown in Figure 1b indicates that both the tetragonal hawleyite [36] and hexagonal greenockite [37] phases are present in approx. 1:1 volume ratio. The energy dispersive X-ray spectroscopy (Figure 1c) indicates the absence of significant impurities, therefore electronic trap states (vide infra) most likely originate from CdS lattice defects. The SEM image (Figure 1d) reveals heavily agglomerated material, for which the particle size statistics were calculated using Image J software [38]. The distribution of crystalline diameters is relatively narrow, ranging from 25 to 110 nm, with an average diameter of 71 ± 3 nm, whereas the distribution maximum is found at 53 ± 1 nm.

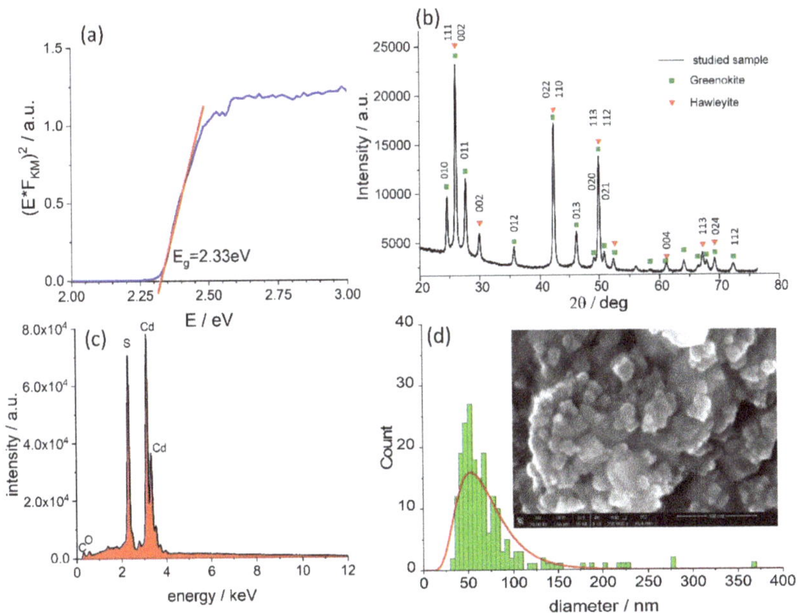

Figure 1. The Tauc plot (**a**), the powder X-ray diffractogram (**b**), the energy-dispersice X-ray spectrum (EDS) (**c**) and the crystallite diameter distribution of the CdS sample (**d**) discussed in this study. Inset shows the SEM image of the studied sample.

2.2. Plasticity of the Artificial Neuron

The composites of cadmium sulfide with multiwalled carbon nanotubes reported in our previous works exhibited memory features that can be functionally associated with the neuronal facilitation (particularly, the paired-pulse facilitation—PPF) in terms of the short-term synaptic plasticity [20]. In the neuroscience, the PPF is considered a neuronal enhancement mechanism which consists of four distinctive processes characterized by different time constants and different physiology [39]. The PPF causes an amplification of the postsynaptic response as a consequence of the increase in stimulating event frequencies at the presynaptic axon [40]. It is believed that the PPF is realized mainly through the accumulation of depolarizing Ca^{2+} ions in the presynaptic neuron [30]. High frequency components of the PPF mechanism (i.e., these characterized by low time constant values) may be useful from the point of view of information processing. These include the fast-decaying facilitation F1 and the slow-decaying facilitation F2 [30,39,41]. The influence of both components manifests itself in the double exponential decay of the postsynaptic response depicted in Figure 2.

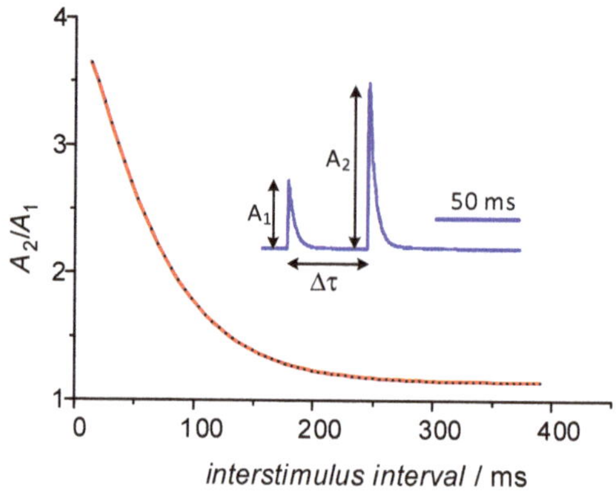

Figure 2. An example of the paired-pulse facilitation observed in nervous system. Adapted from [31].

A similar phenomenon was observed for the photoelectrodes made of nanocrystalline cadmium sulfide (Figure 3). Like in the case of multiwalled carbon nanotubes (MWCNT)-CdScomposite [20], when the interval between light pulses is sufficiently long (over 300 ms for this study) the subsequent photocurrent spikes are unafected by previous states the device was in. However, if the interval between irradiations becomes shorter (e.g., 80 ms) the amplifiation of the second photocurrent response becomes significant. Detailed analysis reveals that the ratio of pulse intensities (the facilitation rate) vs. the time interval between stimuli is best fitted with a biexponential function (1):

$$\frac{A_2}{A_1} = a_1 e^{-\frac{t}{\tau_1}} + a_2 e^{-\frac{t}{\tau_2}} + y_0 \tag{1}$$

which is fully consistent with the previous reports on MWCNT-CdS composite photoelectrodes [20]. The result of the fitting procedure is presented in Figure 3b and the parameters equal to: $a_1 = 0.218 \pm 0.023$, $a_2 = 0.340 \pm 0.014$, $\tau_1 = 19.8 \pm 4.6$ ms, $\tau_2 = 167 \pm 23$ ms and $y_0 = 1.008 \pm 0.013$. The time constants values, which are representative for polycrystalline CdS samples, are slightly higher than those obtained for CdS/MWCNT composites [20]. Interestingly, the determined values are consistent with the parameters typically observed in the case of biological structures [31].

The double exponential decay can be associated with two distinctive trapping/detrapping events characterized by two time constants τ_1 and τ_2. This diversity may originate from the presence of two CdS polymorphs, of which charge trapping states most likely differ. At the same, through the comparison with selected natural learning processes, these two mechanisms may be associated with two components of neuronal plasticity: the fast-decaying facilitation F1 and the slow-decaying facilitation F2. Alternatively, they can be described as manifestations of short- and long-term memory, respectively [42].

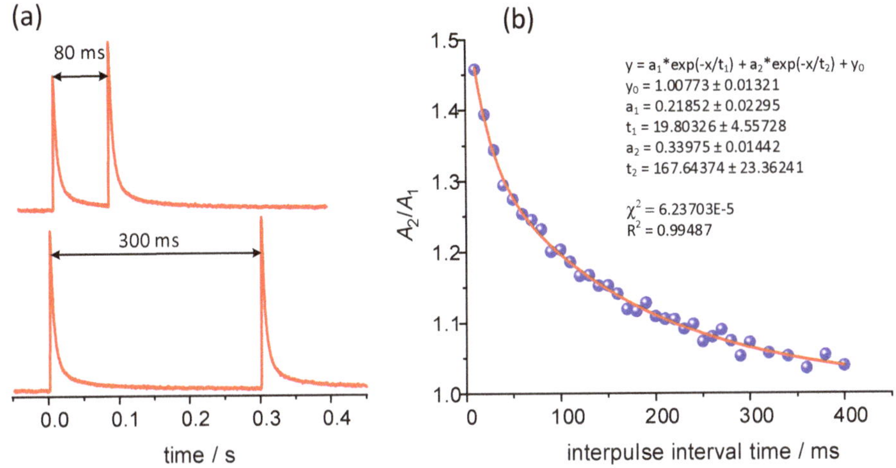

Figure 3. The photocurrent spikes resulting from the pulsed light illumination of CdS-based photoelectrodes (**a**) and the analysis of the photocurrent amplification vs. time interval between subsequent pulses (**b**).

The overall mechanism of photocurrent generation and spikes amplification is summarized in Figure 4. The photoexcitation leads to the electrons transition from the valence to the conduction band (1) and the electron-hole recombination occurs spontaneously afterwards (1′). Electrons in the conduction band can be subsequently transferred through the interface to the conducting substrate (2) and holes can migrate to the surface and react with redox mediators in the electrolyte (2′). At the same time a fraction of electrons from the conduction band becomes trapped within interband states in a very fast process (3). This process efficiently competes with the interfacial electron transfer (2), but once the traps are filled this pathway becomes inactive. The trapped electrons undergo relaxation with the time constants τ_1 (3′) and τ_2 (3″).

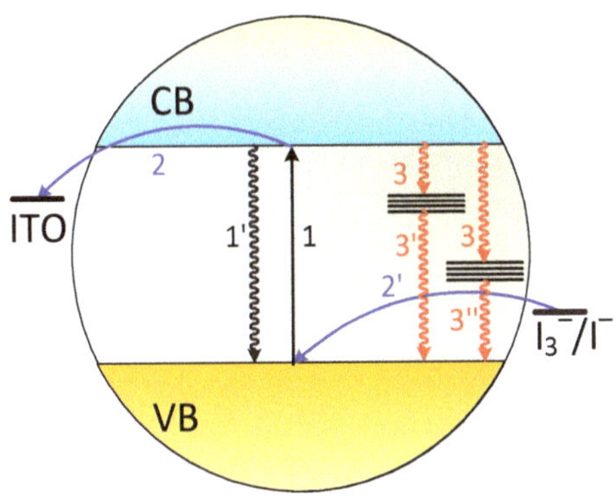

Figure 4. A tentative mechanism of the photocurrent generation and charge carriers trapping in the nanocrystalline CdS sample under consideration.

This simple mechanism provides a platform for the implementation of neuronal dynamics in an artificial, fully inorganic system. Due to its simplicity it can be applied for signal and pattern processing and could be integrated into larger neuromimetic systems. Furthermore, along with bioinspired neuromorphic computing, other information processing paradigms may be implemented within the same system: Boolean logic [43], ternary logic [44] and fuzzy logic [45]. The latter one is especially tempting, as it may contribute to the development of novel neuro-fuzzy information processing devices [46–48].

2.3. Recognition of Digits

A dataset containing 1000 handwritten digits (100 samples of each 0, … , 9 digit) was randomly selected from the MNIST database (Figure 5a). All the images were transformed into binary strings and used for the modulation of a light source. In order to eliminate possible errors resulting from the photoelectrode equilibration or photodegradation, first 20 and last 20 recorded photocurrent profiles were discarded and the remaining 60 patterns were subjected to further processing. First of all, a set of simple classification rules have been developed. These are based on pixel counting, therefore cannot provide a significant separation of the input data. In the first step, each sign (in a form of 28 × 28-pixel image, Figure 5a) was divided into four quadrants labelled $\kappa_1, \ldots, \kappa_4$ (Figure 5b) and the sum of black pixels confined within each quadrant was calculated. In other words, an individual character was associated with a vector $[\Sigma\kappa_1, \Sigma\kappa_2, \Sigma\kappa_3, \Sigma\kappa_4]$ or a point in 4-dimensional space. Subsequently, four 3-dimensional projections were formulated in the following manner: $[\Sigma\kappa_1, \Sigma\kappa_2, \Sigma\kappa_3]$, $[\Sigma\kappa_1, \Sigma\kappa_2, \Sigma\kappa_4]$, $[\Sigma\kappa_1, \Sigma\kappa_3, \Sigma\kappa_4]$ and $[\Sigma\kappa_2, \Sigma\kappa_3, \Sigma\kappa_4]$. For each type of input class (0, … , 9) an ellipsoid with the confidence level of 65% was fitted using 3D Confidence Ellipsoid toolbox in OriginPro 2019.

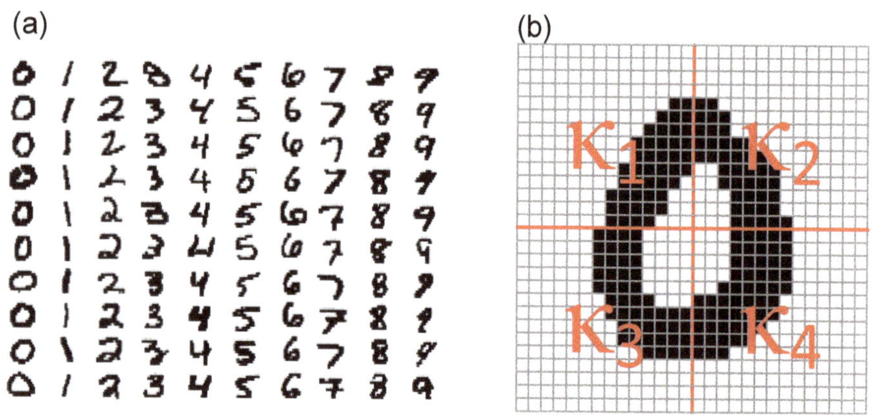

Figure 5. A small sample of the MNIST handwritten digits (**a**) and an image depicting the definition of quadrants for the 28 × 28-pixel image (**b**).

The collection of data points representing all 600 characters under consideration for $[\kappa_1, \kappa_2, \kappa_3]$ combination of quadrants is shown in Figure 6a and the fitted ellipsoids in Figure 6b,c. It can be noticed, that the applied analysis procedure provides a rather poor separation, as the fitted ellipsoids excessively overlap in most of the cases, with the exception of "1" and "9" pair. This result is fully consistent with the initial assumption—simple pixels counting cannot serve as an efficient method for handwritten character recognition.

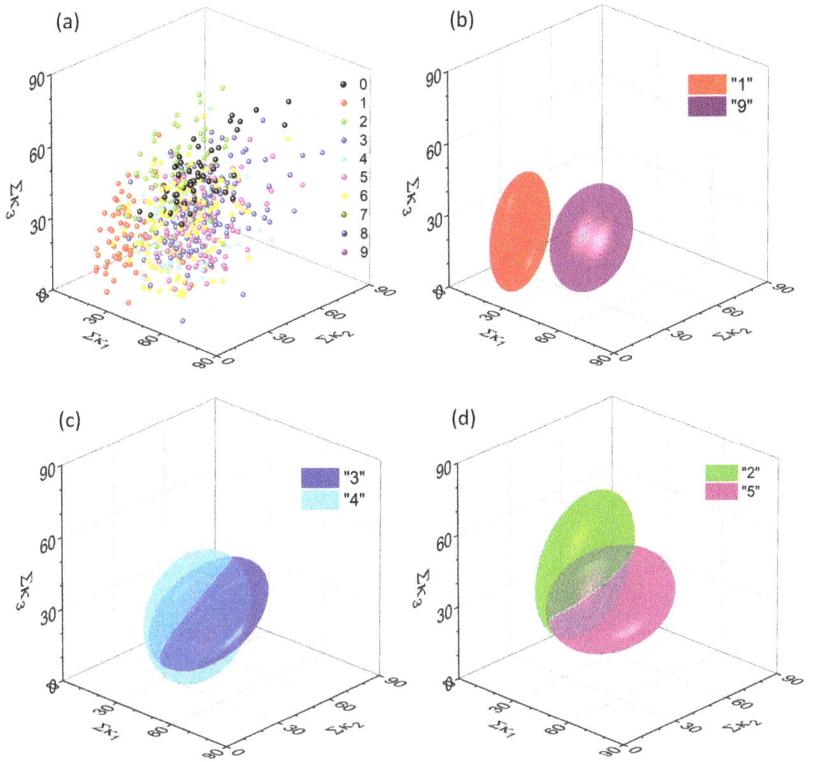

Figure 6. A complete collection of input data points (before feeding them into the single-node neural network) for a set of 600 handwritten digits in the one, arbitrary chosen 3D projection (**a**) and an example of a relatively well-separated pair, which is associated with digits "1" and "9" (**b**). Other ellipsoids overlap significantly, e.g., those for digits "3" and "4" (**c**) or "2" and "5" (**d**).

It can be noticed that only two pairs are completely separated, whereas three others are close to complete separation. Most of these cases concern digit "1" which is substantially different (when the symmetry and number of pixels are taken into account) from any other handwritten digit.

In order to quantify the efficiency of digits recognition in various scenarios, a separability index was defined. Let $V\left(\mathrm{m}\left(\kappa_i, \kappa_j, \kappa_k\right)\right)$ be the volume of an ellipsoid fitted to the digit m representation in $\kappa_i, \kappa_j, \kappa_k$ projection. Then the separability index of the digit m to n will be defined as a ratio between the relative complement of n-ellipsoid in m-ellipsoid to the volume of m-ellipsoid for $m \neq n$ (2):

$$\xi_{m/n} = \frac{V\left(\mathrm{m}\left(\kappa_i, \kappa_j, \kappa_k\right) \backslash \mathrm{n}\left(\kappa_i, \kappa_j, \kappa_k\right)\right)}{V\left(\mathrm{m}\left(\kappa_i, \kappa_j, \kappa_k\right)\right)} \tag{2}$$

The calculated separability indices for the input data in one of the possible projections are collected in Table 1. It is noteworthy that the matrix containing separability indices for a given combination of quadrants is not symmetrical, i.e., $\xi_{m/n} \neq \xi_{n/m}$, since ellipsoids have different volumes. If $m = n$ then $\xi_{m/n} = 0$.

Table 1. The collection of separability indices for the input data in $\kappa_1, \kappa_2, \kappa_3$ projection. The efficiency of data separation is color coded (vide infra) from red (no separation, $\xi = 0$) to green (perfect separation, $\xi = 1$).

	0	1	2	3	4	5	6	7	8	9
0	0.000	1.000	0.753	0.971	0.889	0.812	0.677	0.927	0.838	0.984
1	1.000	0.000	0.963	0.977	0.959	0.996	0.982	0.895	0.999	1.000
2	0.346	0.900	0.000	0.699	0.556	0.563	0.518	0.617	0.569	0.738
3	0.942	0.953	0.770	0.000	0.395	0.360	0.721	0.584	0.523	0.386
4	0.781	0.916	0.668	0.406	0.000	0.505	0.685	0.364	0.558	0.080
5	0.600	0.991	0.649	0.325	0.469	0.000	0.455	0.689	0.355	0.597
6	0.395	0.964	0.659	0.741	0.702	0.520	0.000	0.806	0.692	0.831
7	0.841	0.763	0.684	0.550	0.298	0.680	0.773	0.000	0.694	0.179
8	0.592	0.998	0.590	0.405	0.438	0.236	0.587	0.648	0.000	0.541
9	0.982	1.000	0.890	0.661	0.482	0.789	0.899	0.581	0.797	0.000

In most of the cases the separation is insufficient to allow the unequivocal recognition of handwritten shapes. The exception is a pair {1, 9} for which the separability index is equal to one, as the corresponding ellipsoids do not overlap. It is due to the fact, that "1" differs significantly in terms of pixels distribution between the quadrants. Poor separation can be however greatly improved with the use of even the simplest, the single-node hardware neural network.

In the first step all the characters were converted row-by-row into a stream of bits ("0" for a white pixel and "1" for a black one) and used to modulate the light source according to the scheme presented in Figure 7a,b. The recorded photocurrent spikes (Figure 7c) reflected the sequence of light pulses, but their intensity varied according to the previous states the photoelectrode was in (the short-term memory, vide supra, Figure 3). The photocurrent patterns were subsequently normalized: the amplitude of each signal was divided by the highest intensity recorded for the particular character. The application of various threshold values (Figure 7c,d) acted as a filter for the photocurrent spikes depending on their amplitude. The obtained images with the lowest intensity pixels removed at different thresholds are shown in Figure 8.

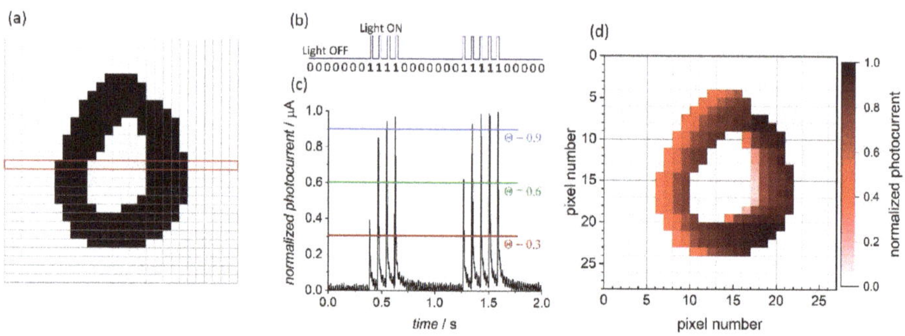

Figure 7. A 28 × 28 pixels image of a handwritten character with a marked row (**a**) translated into a sequence of bits and corresponding light pulses (**b**). A pattern of photocurrent spikes for a given binary input with three thresholds indicated (**c**). An image of the character reconstructed from the normalized photocurrent amplitudes (**d**).

The application of different thresholds (from $\Theta = 0.3$, with virtually no signal filtration to $\Theta = 0.9$, corresponding to the removal of all but the most intense pixels) leads to the evolution of the character image, which depends directly on the neighbors of each particular pixel in the row. Significantly, the "distance" (formerly in space, translated into time intervals between the light pulses) from the closest preceding black pixel determines the weight of the subsequent photocurrent spike amplification. As the result, a simple type of classification according to the scattering of pixels can be achieved. Like in the case of the input data, the output images are subjected to the evaluation of respective separability indices at various threshold values. An example is shown in Table 2.

Figure 9a shows a collection of all data points obtained for one selected projection ($\kappa_1, \kappa_2, \kappa_3$) and one threshold value ($\Theta = 0.7$). They seem to be equally scattered as points for unprocessed data (cf. Figure 6a), but the fitting procedure reveals significant differences. Some ellipsoids, that were initially well separated (e.g., the {1, 9} pair) overlap significantly (Figure 9b). Some other remain unchanged (Figure 9c). More interestingly there are numerous pairs (e.g., {2, 5}, Figure 9d) which are significantly separated upon the data processing with the neuromimetic element.

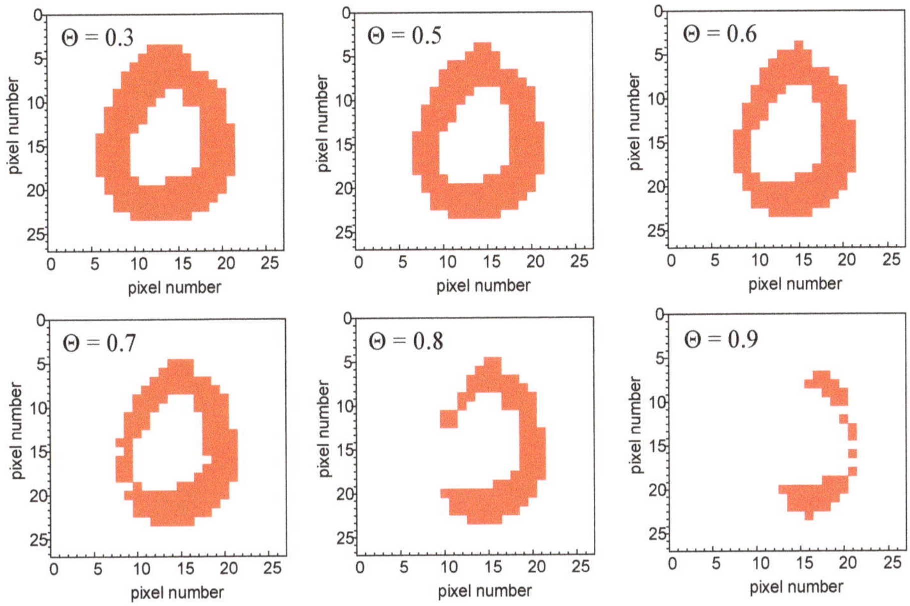

Figure 8. An image of the character from Figure 7 reconstructed from the normalized photocurrent spikes at different threshold values.

Table 2. The collection of separability indices for the output data in $\kappa_1, \kappa_2, \kappa_3$ projection at the threshold $\Theta = 0.7$. The efficiency of data separation is color coded from red (no separation, $\xi = 0$) to green (perfect separation, $\xi = 1$).

	0	1	2	3	4	5	6	7	8	9
0	0.000	0.482	0.783	0.548	0.812	0.452	0.923	0.654	0.782	0.366
1	0.608	0.000	0.936	0.899	0.969	0.376	1.000	0.808	0.984	0.710
2	0.507	0.807	0.000	0.500	0.264	1.000	0.504	0.730	0.653	0.580
3	0.457	0.839	0.735	0.000	0.540	0.964	0.878	0.596	0.611	0.364
4	0.723	0.940	0.522	0.436	0.000	1.000	0.439	0.801	0.672	0.732
5	0.840	0.759	1.000	0.991	1.000	0.000	1.000	0.959	1.000	0.858
6	0.896	1.000	0.705	0.863	0.487	1.000	0.000	0.998	0.854	1.000
7	0.549	0.669	0.845	0.561	0.824	0.817	0.998	0.000	0.788	0.133
8	0.467	0.947	0.626	0.209	0.457	1.000	0.735	0.603	0.000	0.512
9	0.670	0.801	0.904	0.725	0.906	0.747	1.000	0.654	0.896	0.000

Upon data treatment with the neuromimetic element the separability is significantly improved. Six pairs of digits are completely separated and two others are close to complete separation. Furthermore, these pairs are different than those separated with the use of the pixel counting method (cf. Table 1).

The improvement of digits classification can be visually evaluated by the comparison of color-coded Tables 1 and 2. In order to perform a global quantitative evaluation of the separation efficiencies and to assess the improvement associated with the use of the single-node hardware neural device, a separability ratio ($\Xi_{m/n}$) was defined as a ratio of the separability index calculated for the processed data to the separability index determined for the input data (for $m \neq n$) (3):

$$\Xi_{m/n} = \frac{\xi_{m/n}^{output}}{\xi_{m/n}^{input}} \quad (3)$$

and $\Xi_{m/n} = 0$ for $m = n$. Detailed analysis of calculated values provides information on the discussed procedure efficiency even in the case of significantly overlapped ellipsoids. Selected separability ratios for $\kappa_1, \kappa_2, \kappa_3$ projection at the threshold $\Theta = 0.7$ are presented in Table 3.

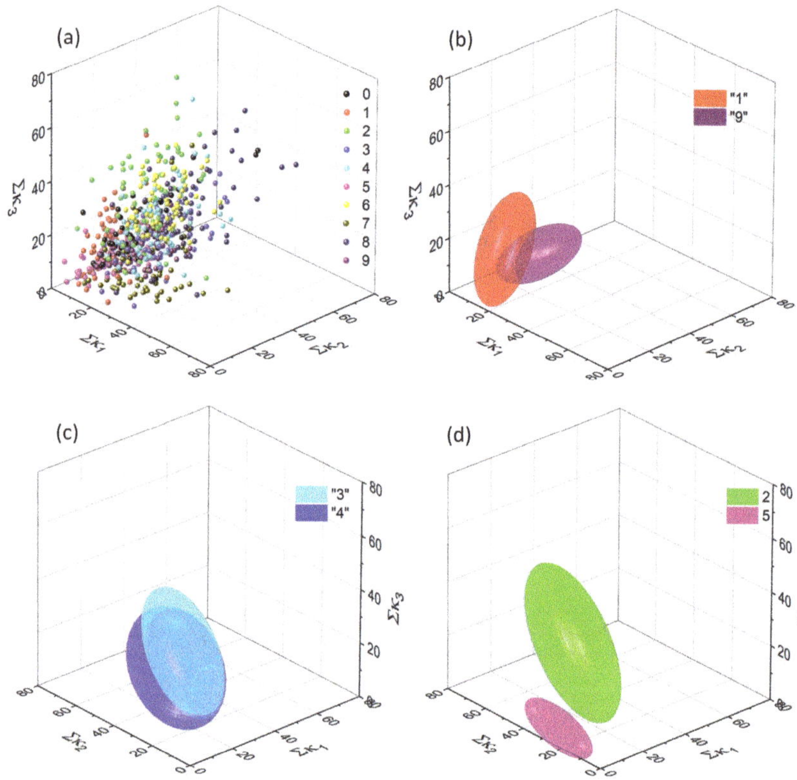

Figure 9. A complete collection of output data points (after feeding them into the single-node neural network) for a set of 600 handwritten digits in the one, arbitrary chosen 3D projection (corresponding to the one shown in Figure 6) (**a**) and an example of significantly overlapped ellipsoids, corresponding to digits "1" and "9" (**b**), "3" and "4" (**c**). The majority of other ellipsoids are separated better, than in the case of untreated data—e.g., those associated with digits "2" and "5" (**d**).

Table 3. The collection of separability ratios ($\Xi_{m/n}$) for the output data in $\kappa_1, \kappa_2, \kappa_3$ projection at the threshold $\Theta = 0.7$. The efficiency of data separation is color coded from red—significantly decreased separability ($\Xi < 0.5$), through yellow—slightly decreased separability ($0.5 < \Xi < 1$), green—moderately improved separability ($1 < \Xi < 1.5$), blue—significantly improved separability ($1.5 < \Xi < 2$) to navy blue—outstanding improvement of separability ($\Xi > 2$).

	0	1	2	3	4	5	6	7	8	9
0	0.000	0.482	1.040	0.564	0.913	0.556	1.363	0.705	0.933	0.372
1	0.608	0.000	0.971	0.919	1.011	0.377	1.019	0.903	0.984	0.710
2	1.468	0.897	0.000	0.715	0.475	1.777	0.974	1.183	1.147	0.787
3	0.485	0.881	0.954	0.000	1.369	2.682	1.219	1.019	1.167	0.945
4	0.926	1.026	0.781	1.074	0.000	1.979	0.642	2.203	1.205	9.127
5	1.398	0.766	1.542	3.055	2.134	0.000	2.198	1.393	2.818	1.438
6	2.267	1.037	1.070	1.166	0.694	1.923	0.000	1.239	1.233	1.204
7	0.652	0.877	1.235	1.021	2.770	1.202	1.291	0.000	1.135	0.742
8	0.788	0.949	1.061	0.515	1.044	4.229	1.253	0.931	0.000	0.947
9	0.682	0.801	1.016	1.097	1.877	0.947	1.112	1.125	1.125	0.000

It can be noticed that the overall improvement of separability is achieved with at least twofold increase of the index value for 10 pairs compared to the twofold decrease in five instances. A similar situation is also observed for other projections at this threshold. This qualitative picture suggests an improvement of handwritten digits recognition upon application of a neuromimetic element in data processing. A quantitative estimation of classification improvement can be obtained through simple numerical analysis of output data. Due to the analysis complexity (four quadrant combinations for eight different threshold values), various separation scenarios (depending on the chosen threshold and projection) are possible. Their overall efficiency can be evaluated using an integral separability index Ω, which acts as a global parameter indicating performance of the system for all of the above-mentioned variables. For each separation scenario it can be defined as follows (4):

$$\Omega\left(\kappa_i, \kappa_j, \kappa_k, \Theta\right) = \frac{\sum\limits_m \sum\limits_n \xi_{m/n}^{output}}{\sum\limits_m \sum\limits_n \xi_{m/n}^{input}}, m \neq n \tag{4}$$

The above-mentioned dependency of the Ω function values is depicted in Figure 10. It can be noted, that in the majority of investigated separation scenarios the integral separability index after the treatment with the single-node hardware neural network is significantly higher than the value calculated for the unprocessed input data. In three cases the selection of a low threshold value (a situation which results in an insufficient filtration of pixels due to inadequate exploitation of memory features) leads to the inferior separation. On the other hand, when the short-term memory of the system is optimally utilized, the recognition of handwritten digits increases. For two quadrant combinations an optimal threshold value exists, which is fully consistent with the expectations—too deep filtration removes too many pixels and all the data points (cf. Figures 6a and 9a) interfuse at the origin of the coordinate system.

3. Conclusions

Surprisingly, even a primitive hardware realization of the neural network architecture based on a single-node exhibiting the short-term memory can significantly improve pattern recognition. Classification based solely on the number of black pixels encompassed by each of the four quadrants the character image is divided into (cf. Figure 5b) is insufficient—only a few characters of a specific symmetry (e.g., "1") could be distinguished using this primitive procedure. The application of the optoelectronic element with PPF functionality enhances tremendously (for such a simple device) the classification efficiency. This change is based on the extraction of a new feature of the studied data—the scattering of pixels within the 28 × 28 matrix. The high dispersion leads to the negligible photocurrent amplification. On the contrary, the digits with large groups of pixels are characterized by a higher number of counts for relatively high thresholds, for which the short-term memory of the system and resulting photocurrent amplification is strongly pronounced.

The system presented in this work is a single node neural device, the operation of which is based on the unsupervised learning paradigm involving the short-term memory of the device. The simple pixel counting method gives precise information on the size of the characters (therefore digit "1" separates well in all of the cases) and indirectly on their symmetry (it can be achieved by an appropriate selection of quadrant combinations). Application of a neuromimetic element allows further information processing, particularly the extraction of information on pixel "agglomeration", at least at the single row level. This process is analogous to Sammon mapping [49], but does not involve the reduction of data space dimensionality (Figure 11).

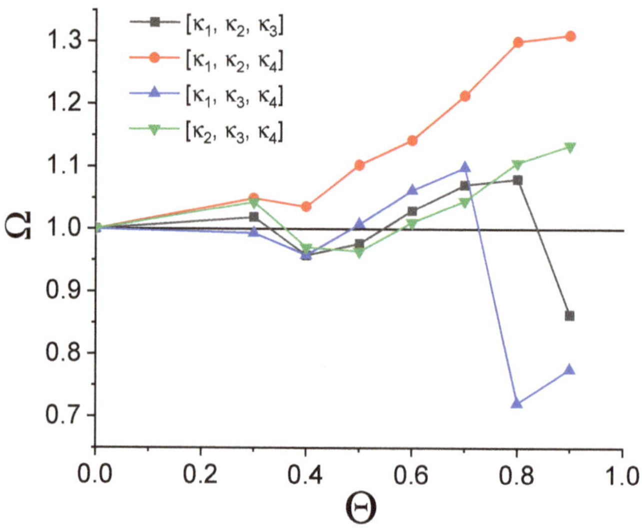

Figure 10. The dependence of the integral separability index Ω vs. the threshold value Θ. The horizontal line indicates the integral separability values determined for the input data. Threshold $\Theta = 0$ corresponds to unprocessed input data.

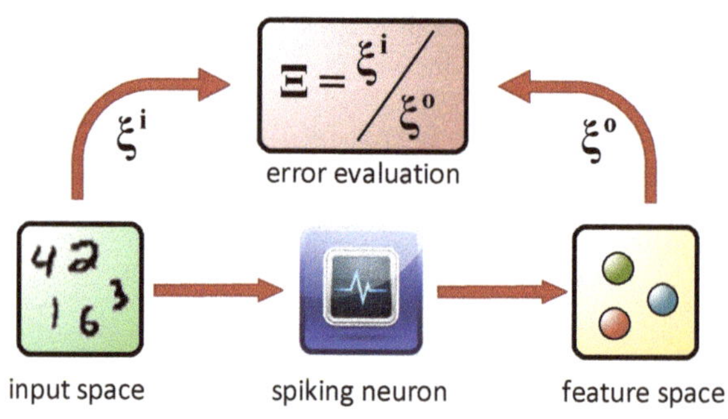

Figure 11. A diagram depicting the data flow and the efficiency evaluation of the unsupervised classification system under consideration.

The presented optoelectronic single-node neural device is superior compared to software implementations in terms of the error resistance and the energy efficiency. These features of the discussed system make it a potential low-cost pre-processing unit. Furthermore, due to the operation based on time-series and the intrinsic short-term memory, it can be combined with selected aspects of the reservoir computing paradigm—in one of the possible scenarios, where a delayed feedback loop is used, the virtual neurons could significantly affect the system performance. At the same time, the analog character of the system allows the implementation of the fuzzy logic system, yielding a new class of hardware neuro-fuzzy devices.

The research presented in this paper supports the concept of heterodic computation [27,28]. It clearly shown that the performance of a simple numerical algorithm (classification based on pixel counting) can be improved by in-materio computational component, which itself cannot perform any classification tasks.

4. Materials and Methods

Commercially available cadmium sulfide (POCH, Las Condes, Chile), potassium nitrate (Avantor, Radnor, PA, USA), potassium iodide (Aldrich, St. Louis, MO, USA) and iodine (Aldrich, St. Louis, MO, USA) were used as received.

Working electrodes were prepared from polyethylene terephthalate (PET, Camarillo, CA, USA) foil coated with indium tin oxide (Aldrich, St. Louis, MO, USA). These substrates were washed carefully with diluted detergent solution, deionized water and isopropanol. They were then dried in air. The cadmium sulfide was ground with deionized water in an agate mortar to a thick paste and deposited onto the freshly cleaned substrates using a screen-printing machine (MikMetal, Masis St, Yerevan, Armenia) equipped with a 80 mesh polymer grid.

The UV/Vis spectra were recorded using Lambda 750 spectrophotometer (Perkin Elmer, Waltham, MA, USA) within the wavelength range of 200–2000 nm. Barium sulfate of spectral purity was used as a reference material. The X–ray diffraction patterns were recorded with Empyrean (Cu $\lambda_{K\alpha 1}$ = 1.54060 Å) diffractometer (PANalytical, Malvern, UK) at room temperature with 2θ values ranging from 20 to 80 degrees. The scanning electron images were taken on Versa 3d (FEI, Lausanne, Switzerland) scanning electron microscope operating at 20 kV with an Everhart-Thornley detector. The chemical composition of the CdS sample was confirmed using the energy dispersive X-ray spectroscopy. All electrochemical measurements were performed with the use of SP-300 potentiostat (BioLogic, Cary, NC, USA). Luxeon Star/O Royal Blue diode (465 nm, the total radiometric power of 110 mW) was used as the light source. It was powered through the WA-301 wideband amplifier (Aim-TTI, Cambridgeshire, England). Pulse sequences were generated with TG2512A arbitrary function generator (Aim-TTI, Cambridgeshire, England) triggered with Arduino Uno R3 system.

The photoelectrochemical experiments were performed in air-equilibrated electrolytes using a three-electrode configuration. As the photoactive component the screen-printed CdS electrodes, immersed in an aqueous electrolyte containing 0.1 M KNO_3, 0.001 M KI and 0.0001 M I_2, were used. A saturated Ag/AgCl electrode was used as a reference electrode and a platinum wire as a counter electrode. The positive voltage (400 mV) was applied to the working electrode and the photocell was irradiated with short light pulses (300 µs).

The experiment automation was realized based on the program written in Arduino C++ language. All the necessary data processing was performed with the use of programs written in Python 3.7.2.

Author Contributions: M.L., D.P. and K.P. have performed all photoelectrochemical experiments. D.P. and K.P. have selected appropriate data sets. K.P. and K.S. have developed algorithms for data processing and designed the experimental setup. K.P. wrote the software necessary for data handling and measurements automation. All the authors contributed to the data analysis and the manuscript preparation.

Funding: This research was funded by National Science Centre (Poland) within the MAESTRO project, contract No. UMO-2015/18/A/ST4/00058. D.P. acknowledges partial support from European Union within the EU Project POWR.03.02.00-00-I004/16. K.P. was supported by the Foundation for Polish Science (FNP) and acknowledges partial support from National Science Centre within the PRELUDIUM project, contract No. UMO-2016/21/N/ST3/00469.

Acknowledgments: The authors thank Grzegorz Cios and Marianna Marciszko for the XRD measurements and Kapela Pilaka for countless discussions of an extremely fruitful character on the design of hardware neural networks.

Conflicts of Interest: The authors declare no conflict of interest.

References

1. Perez, L.; Wang, J. The effectiveness of data augmentation in image classification using deep learning. *arXiv* **2017**, arXiv:1712.04621 2017.
2. Basheer, I.A.; Hajmeer, M. Artificial neural networks: fundamentals, computing, design, and application. *J. Microbiol. Methods* **2000**, *43*, 3–31. [CrossRef]
3. Querlioz, D.; Bichler, O.; Gamrat, C. Simulation of a memristor-based spiking neural network immune to device variations. In Proceedings of the 2011 International Joint Conference on Neural Networks, San Jose, CA, USA, 31 July–5 August 2011; pp. 1775–1781.
4. Liu, W.; Wang, Z.; Liu, X.; Zeng, N.; Liu, Y.; Alsaadi, F.E. A survey of deep neural network architectures and their applications. *Neurocomputing* **2017**, *234*, 11–26. [CrossRef]
5. Nguyen, A.; Yosinski, J.; Clune, J. Deep neural networks are easily fooled: High confidence predictions for unrecognizable images. In Proceedings of the IEEE Conference on Computer Vision and Pattern Recognition, Boston, MA, USA, 7–12 June 2015; pp. 427–436.
6. Su, J.; Vargas, D.V.; Sakurai, K. One pixel attack for fooling deep neural networks. *IEEE Trans. Evol. Comput.* **2019**. [CrossRef]
7. Nawrocki, R.A.; Voyles, R.M.; Shaheen, S.E. A mini review of neuromorphic architectures and implementations. *IEEE Trans. Electron. Dev.* **2016**, *63*, 3819–3829. [CrossRef]
8. Akopyan, F.; Sawada, J.; Cassidy, A.; Alvarez-Icaza, R.; Arthur, J.; Merolla, P.; Imam, N.; Nakamura, Y.; Datta, P.; Nam, G.-J. Truenorth: Design and tool flow of a 65 mw 1 million neuron programmable neurosynaptic chip. *IEEE Trans. Computer-Aided Des. Integr. Circ. Syst.* **2015**, *34*, 1537–1557. [CrossRef]
9. Wu, Y.; Yu, S.; Wong, H.-S.P.; Chen, Y.-S.; Lee, H.-Y.; Wang, S.-M.; Gu, P.-Y.; Chen, F.; Tsai, M.-J. AlOx-based resistive switching device with gradual resistance modulation for neuromorphic device application. In Proceedings of the 4th IEEE International Memory Workshop, Milan, Italy, 20–23 May 2012; pp. 1–4.
10. Jang, J.-W.; Park, S.; Burr, G.W.; Hwang, H.; Jeong, Y.-H. Optimization of conductance change in Pr 1–x Ca x MnO 3-based synaptic devices for neuromorphic systems. *IEEE Electron Dev. Lett.* **2015**, *36*, 457–459. [CrossRef]
11. Yu, S.; Wu, Y.; Jeyasingh, R.; Kuzum, D.; Wong, H.-S.P. An electronic synapse device based on metal oxide resistive switching memory for neuromorphic computation. *IEEE Trans. Electron. Dev.* **2011**, *58*, 2729–2737. [CrossRef]
12. Jin, X.; Rast, A.; Galluppi, F.; Davies, S.; Furber, S. Implementing spike-timing-dependent plasticity on SpiNNaker neuromorphic hardware. In Proceedings of the 2010 International Joint Conference on Neural Networks (IJCNN), Barcelona, Spain, 18–23 July 2010; pp. 1–8.
13. Qiao, N.; Mostafa, H.; Corradi, F.; Osswald, M.; Stefanini, F.; Sumislawska, D.; Indiveri, G. A reconfigurable on-line learning spiking neuromorphic processor comprising 256 neurons and 128K synapses. *Front. Neurosci.* **2015**, *9*, 141. [CrossRef]
14. Indiveri, G.; Linares-Barranco, B.; Legenstein, R.; Deligeorgis, G.; Prodromakis, T. Integration of nanoscale memristor synapses in neuromorphic computing architectures. *Nanotechnology* **2013**, *24*, 384010. [CrossRef]
15. Wang, J.; Hu, S.; Zhan, X.; Yu, Q.; Liu, Z.; Chen, T.P.; Yin, Y.; Hosaka, S.; Liu, Y. Handwritten-digit recognition by hybrid convolutional neural network based on hfo 2 memristive spiking-neuron. *Sci. Rep.* **2018**, *8*, 12546. [CrossRef]
16. Gentili, P.L.; Giubila, M.S.; Germani, R.; Romani, A.; Nicoziani, A.; Spalletti, A.; Heron, B.M. Optical Communication among Oscillatory Reactions and Photo-Excitable Systems: UV and Visible Radiation Can Synchronize Artificial Neuron Models. *Angew. Chem. Int. Ed.* **2017**, *56*, 7535–7540. [CrossRef] [PubMed]
17. Szaciłowski, K. Molecular Logic Gates Based on Pentacyanoferrate Complexes: From Simple Gates to Three-Dimensional Logic Systems. *Chem. Eur. J.* **2004**, *10*, 2520–2528. [CrossRef] [PubMed]
18. Mech, J.; Kowalik, R.; Podborska, A.; Kwolek, P.; Szaciłowski, K. Arithmetic device based on multiple Schottky-like junctions. *Aust. J. Chem.* **2010**, *63*, 1330–1333. [CrossRef]

19. Gentili, P.L.; Giubila, M.S.; Germani, R.; Heron, B.M. Photochromic and luminescent compounds as artificial neuron models. *Dyes Pigments* **2018**, *156*, 149–159. [CrossRef]
20. Pilarczyk, K.; Podborska, A.; Lis, M.; Kawa, M.; Migdał, D.; Szaciłowski, K. Synaptic Behavior in an Optoelectronic Device Based on Semiconductor-Nanotube Hybrid. *Adv. Electr. Mater.* **2016**, 1500471. [CrossRef]
21. Pilarczyk, K.; Wlaźlak, E.; Przyczyna, D.; Blachecki, A.; Podborska, A.; Anathasiou, V.; Konkoli, Z.; Szaciłowsk, K. Molecules, semiconductors, light and information: Towards future sensing and computing paradigms. *Coord. Chem. Rev.* **2018**, *365*, 23–40. [CrossRef]
22. Pilarczyk, K.; Daly, B.; Podborska, A.; Kwolek, P.; Silverson, V.A.D.; de Silva, A.P.; Szaciłowski, K. Coordination chemistry for information acquisition and processing. *Coord. Chem. Rev.* **2016**, *325*, 135–160. [CrossRef]
23. Horsman, C.; Stepney, S.; Wagner, R.C.; Kendon, V. When does a physical system compute? *Proc. Royal. Soc. A* **2014**, *470*, 20140182. [CrossRef]
24. Stepney, S. The neglected pillar of material computation. *Physica D* **2008**, *237*, 1157–1164. [CrossRef]
25. Stepney, S.; Rasmussen, S.; Amos, M. (Eds.) *Computational Matter*; Springer: Cham, Switzerland, 2018.
26. Szaciłowski, K. *Infochemistry. Information Processing at the Nanoscale*; John Wiley & Sons: Chichester, UK, 2012.
27. Kendon, V.; Sebald, A.; Stepney, S. Heterotic computing: exploiting hybrid computational devices. *Phil. Trans. Royal Soc. A* **2015**, *373*, 20150091. [CrossRef] [PubMed]
28. Kendon, V.; Sebald, A.; Stepney, S. Heterotic computing: past, present and future. *Phil. Trans. Royal Soc. A* **2015**, *373*, 20140225. [CrossRef] [PubMed]
29. LeCun, Y.; Bottou, L.; Bengio, Y.; Haffner, P. Gradient-based learning applied to document recognition. *Proc. IEEE* **1998**, *86*, 2278–2324. [CrossRef]
30. Zucker, R.S.; Regehr, W.G. Short-term synaptic plasticity. *Annu. Rev. Physiol.* **2002**, *64*, 355–405. [CrossRef] [PubMed]
31. Jackman, S.L.; Regehr, W.G. The mechanisms and functions of synaptic facilitation. *Neuron* **2017**, *94*, 447–464. [CrossRef] [PubMed]
32. Appeltant, L.; Soriano, M.C.; Van der Sande, G.; Danckaert, J.; Massar, S.; Dambre, J.; Schrauwen, B.; Mirasso, C.R.; Fischer, I. Information processing using a single dynamical node as complex system. *Nat. Commun.* **2011**, *2*, 468. [CrossRef] [PubMed]
33. Zelaya-Angel, O.; Alvarado-Gil, J.J.; Lozada-Morales, R.; Vargas, H.; Fereira da Silva, A. Bandgap shift in CdS semiconductor by photoacoustic spectroscopy: evidence of a cubic to hexagonal lattice transition. *Appl. Phys. Lett.* **1994**, *64*, 291–293. [CrossRef]
34. Degen, T.; Sadki, M.; Bron, E.; König, U.; Nénert, G. The HighScore suite. *Powder Diffr.* **2014**, *29*, S13–S18. [CrossRef]
35. Rietveld, H.M. A profile ferinement method for nuclear and magnetic structures. *J. Appl. Cryst.* **1969**, *2*, 65–71. [CrossRef]
36. Skinner, B.J. Unit-cell edges of natural and synthetis sphalerites. *Am. Miner.* **1961**, *46*, 1399–1411.
37. Xu, Y.N.; Ching, W.Y. Electronic, optical and structural properties of some wurzite crystals. *Phys. Rev. B* **1993**, *48*, 4335–4351. [CrossRef] [PubMed]
38. Rasband, W.S. *ImageJ*; U.S. National Institutes of Health: Bethesda, MD, USA, 2012.
39. Fisher, S.A.; Fischer, T.M.; Carew, T.J. Multiple overlapping processes underlying short-term synaptic enhancement. *Trends Nneurosci.* **1997**, *20*, 170–177. [CrossRef]
40. Salin, P.A.; Scanziani, M.; Malenka, R.C.; Nicoll, R.A. Distinct short-term plasticity at two excitatory synapses in the hippocampus. *Proc. Natl. Acad. Sci. USA* **1996**, *93*, 13304–13309. [CrossRef] [PubMed]
41. Wang, Z.; Joshi, S.; Savel'ev, S.E.; Jiang, H.; Midya, R.; Lin, P.; Hu, M.; Ge, N.; Strachan, J.P.; Li, Z. Memristors with diffusive dynamics as synaptic emulators for neuromorphic computing. *Nat. Mater.* **2017**, *16*, 101. [CrossRef] [PubMed]
42. Yu, Q.; Tang, H.; Hu, J.; Tan, K.C. *Neuromorphic Cognitive Systems. A Learning and Memory Centered Approach*; Springer International Publishing AG: Cham, Switzerland, 2017.
43. Szaciłowski, K.; Macyk, W.; Stochel, G. Light-driven OR and XOR programmable chemical logic gates. *J. Am. Chem. Soc.* **2006**, *128*, 4550–4551. [CrossRef] [PubMed]
44. Warzecha, M.; Oszajca, M.; Pilarczyk, K.; Szaciłowski, K. A three-valued photoelectrochemical logic device realising accept anything and consensus operations. *Chem. Commun.* **2015**, *51*, 3559–3561. [CrossRef] [PubMed]

45. Blachecki, A.; Mech-Piskorz, J.; Gajewska, M.; Mech, K.; Pilarczyk, K.; Szaciłowski, K. Organotitania-Based Nanostructures as a Suitable Platform for the Implementation of Binary, Ternary, and Fuzzy Logic Systems. *ChemPhysChem* **2017**, *18*, 1798–1810. [CrossRef]
46. Kasabov, N. From multilayer perceptrons and neurofuzzy systems to deep learning machines: which method to use? A surver. *Int. J. Inf. Technol. Security* **2017**, *9*, 3–24.
47. Kar, S.; Das, S.; Gosh, P.K. Applications of neuro fuzzy systems: A brief review and future outline. *Appl. Soft Comput.* **2014**, *15*, 243–259. [CrossRef]
48. Bavandpour, M.; Bagheri-Shouraki, S.; Soleimani, H.; Ahmadi, A.; Linares-Barranco, B. Spiking neuro-fuzzy clustering system and its memristo crossbar basd implementation. *Microelectron. J.* **2014**, *45*, 1450–1462. [CrossRef]
49. Sammon, J.W. A nonlinear mapping for data strucutre analysis. *IEEE Trans. Computers* **1969**, *C18*, 401–409. [CrossRef]

Sample Availability: Samples of the polycrystalline CdS are available from the authors on request. The full set of output data with separability indices for all possible classification scenarios is available on request. The programs used for the data processing are available on request.

molecules

MDPI

Perspective

Towards a Stochastic Paradigm: From Fuzzy Ensembles to Cellular Functions

Monika Fuxreiter *

MTA-DE Laboratory of Protein Dynamics, Department of Biochemistry and Molecular Biology, H-4032 Debrecen, Hungary

Academic Editor: Pier Luigi Gentili
Received: 21 October 2018; Accepted: 16 November 2018; Published: 17 November 2018

Abstract: The deterministic sequence → structure → function relationship is not applicable to describe how proteins dynamically adapt to different cellular conditions. A stochastic model is required to capture functional promiscuity, redundant sequence motifs, dynamic interactions, or conformational heterogeneity, which facilitate the decision-making in regulatory processes, ranging from enzymes to membraneless cellular compartments. The fuzzy set theory offers a quantitative framework to address these problems. The fuzzy formalism allows the simultaneous involvement of proteins in multiple activities, the degree of which is given by the corresponding memberships. Adaptation is described via a fuzzy inference system, which relates heterogeneous conformational ensembles to different biological activities. Sequence redundancies (e.g., tandem motifs) can also be treated by fuzzy sets to characterize structural transitions affecting the heterogeneous interaction patterns (e.g., pathological fibrillization of stress granules). The proposed framework can provide quantitative protein models, under stochastic cellular conditions.

Keywords: protein dynamics; conformational heterogeneity; promiscuity; fuzzy complexes; higher-order structures; protein evolution; fuzzy set theory; artificial intelligence

---

1. The Structure-Function Paradigm

Protein functions take place in space and time. Structure-function principles, however, relate a protein sequence to biological activity, only via the spatial coordinates of the residues [1,2]:

$$\text{SEQUENCE} \rightarrow \underset{(x,y,z)}{\text{STRUCTURE}} \rightarrow \underset{(x,y,z,t)}{\text{FUNCTION}} \tag{1}$$

The three-dimensional organization of amino acids brings different chemical groups into proximity [3,4], creating specific microenvironments for biological activities. The emerging active sites, for example, can catalyze chemical reactions at significantly faster rates, than in solution [5,6]. The classical, deterministic Paradigm 1 establishes an unambiguous connection between the protein sequence and its function, via a unique structure.

2. The Ensemble View

The energy landscapes of proteins are, in reality, more complicated. Proteins fluctuate among various conformations ('macrostates') and sub-states ('microstates'), which need to be considered for their relevant functioning [7–9]. A wide spectrum of dynamical transitions [10,11]—from local movements (e.g., sidechain rotations [12]) to large-amplitude collective motions (e.g., domain repositioning [13])—generates conformational ensembles, which, however, are not trivial to link to the function. How can structure-function relationships account for protein dynamics? If protein structure is described as an ensemble, the populations of the relevant sub-states, as well as the rate of interconversion between them, must be experimentally determined for each biological activity.

$$\begin{array}{ccccc} \text{SEQUENCE} \rightarrow & \text{CONFORMATIONAL ENSEMBLE} & \rightarrow & \text{FUNCTION} & \\ & \{p_{cs_1}, \ldots, p_{cs_N}\} & & (x,y,z,t) & \quad (2) \\ & \{k_{cs_1}, \ldots, k_{cs_N}\} & & & \end{array}$$

where p_{CS_i} is the probability of the given conformational sub-state CS_i, N is the number of sub-states, and $\{k_{CS_i}\}$ is the set of rates, corresponding to the conversions between $CS_i \rightarrow CS_j$, where j corresponds to all the other sub-states. Even if the number of conformational states is reduced to a few functionally relevant ones, characterizing both their thermodynamic and kinetic properties is a daunting task [14,15]. Furthermore, the deterministic relationship between the ensemble parameters and a unique function is also influenced by the environmental conditions.

3. Adaptation to Stochastic Cellular Conditions

Proteins function under rapidly changing extracellular signals and intracellular milieu, which is shaped by cellular diffusion and transport, stochastic gene expression, degradation, and other environmental fluctuations. These factors present stochastic conditions for protein evolution [16–18] leading to 'noise' in biological innovations [19], which is reflected by redundancies and ambiguities in sequences [20], structures [21], and functions [22]. On the one hand, proteins attempt to minimize functional noise. For example, higher-order structures emerge to reduce noise-to-signal ratio for low-affinity substrates [23–25]. On the other hand, ambiguities and redundancies in sequence, structure and function facilitate dynamic adaptation [26]. Proteins evolve under these two opposing constraints to optimize fitness under given cellular conditions.

4. Ambiguity and Redundancy in Sequence, Structure, and Function

The re-formatted paradigm (2), still implies that a given sequence generates a well-defined ensemble, which belongs to a specific function. The stochastic cellular conditions lead to the following observations, which violate the classical paradigm: (i) A considerable proportion of proteins exhibit multiple, simultaneous activities, often referred to as promiscuity or moonlighting [27]. (ii) Certain biological activities (i.e., signaling) are related to heterogeneous conformational ensembles, which are mixtures of different functional ensembles [28]. (iii) Some proteins exhibit a weak sequence dependence, i.e., a large degree of tolerance towards sequence modifications [29]. These observations stem from redundancies in sequence or structure, coupled to ambiguities in function. The same ensemble may perform multiple functions (*functional promiscuity*); the same sequence may be organized into multiple functional ensembles, depending on the context (*conformation and interaction heterogeneity*); and multiple sequences may encode the same conformational ensemble (*sequence redundancy*). These problems, which reflect a more complex relationship between the sequence, structure, and function of proteins, are detailed below.

5. Functional Promiscuity

Metabolic enzymes often catalyze reactions on the non-canonical substrates, some of which are also relevant physiologically [27,30,31]. Functional promiscuity may parallel organism complexity [32], or be driven by network context [33]. Promiscuous activities can serve as starting points to engineer new enzymes [34]. Tailored selection pressures may optimize latent activities to overcome the primary function by >10^9-fold [35]. Functional transitions are usually initiated by 'neutral drifts', with a negligible impact on the original activity [36,37]. That is, the optimization of a promiscuous function initially exploits the inherent variations in structure [38] and dynamics [39]. Functional transition of a phosphotriesterase to arylesterase [35], for example, is coupled to increasing structural divergence between the two subunits, until the two activities become comparable (Figure 1A).

In contrast, specialization for the new activity is accompanied by structural convergence (Figure 1A). Similarly, 'freezing' out unnecessary motions offers another route to optimize enzymatic efficiency [6]. Along these lines, principal modes derived from structure [40] often presage or follow the evolutionary changes [41,42].

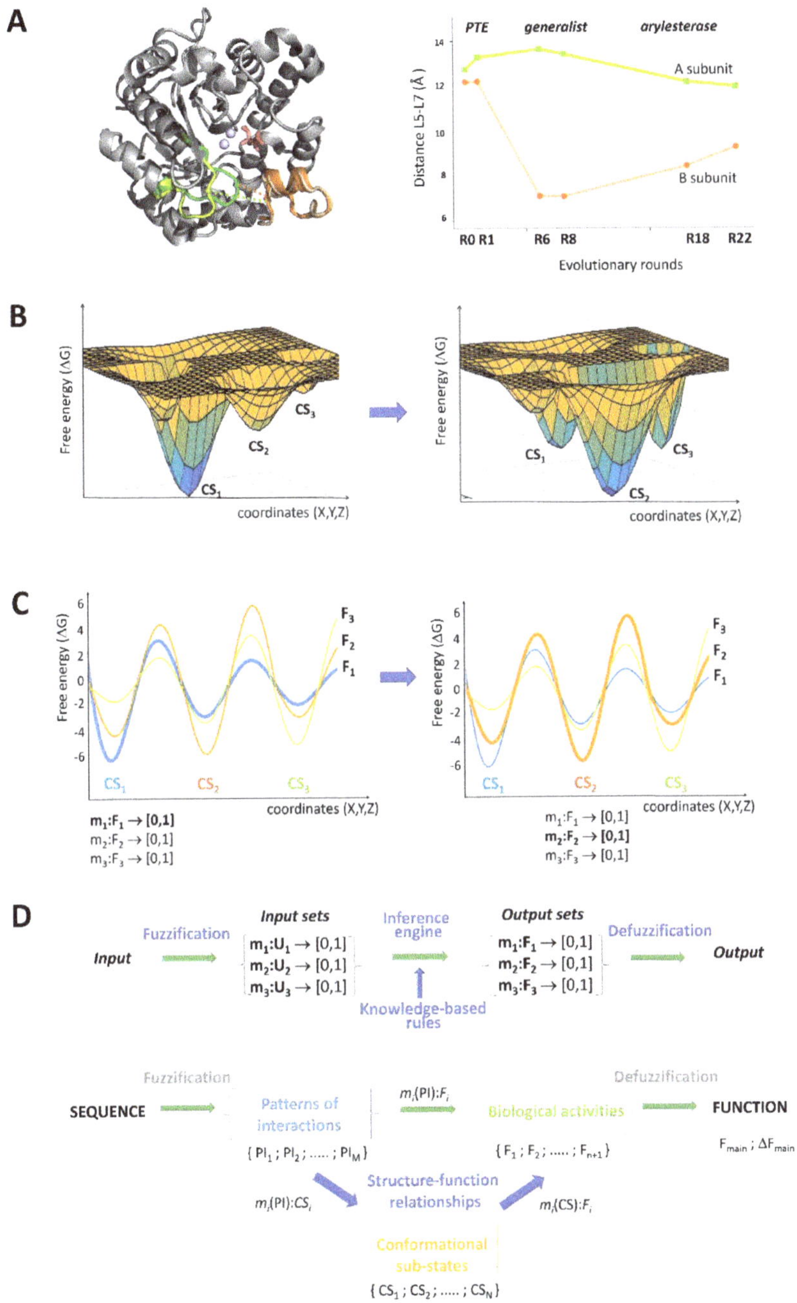

Figure 1. Towards a stochastic structure-function relationship. (**A**) Structural diversity increases with functional promiscuity. The distance between the L5 (*lime, green*) and L7 (*wheat, orange*) loops (A204 C–G273 C) deviates in the two subunits (*superimposed*) of a dimeric phosphotriesterase (PTE) enzyme (PDB code: 4xag [39]). During laboratory evolution into arylesterase, the structural difference increases as the two activities become comparable (R1 → R6), while it decreases during specialization (R8 → R22). (**B**) Free energy landscape changes upon adaptation of proteins. Functional alterations shift the relative populations of conformational sub-states, but may not impact the ruggedness of the landscape. (**C**) Conformational sub-states (CSs) contribute to multiple free landscapes. The functional noise (uncertainty of F_1, F_2, F_3) of the main activity (*bold*) can be quantified by fuzzy membership functions. (**D**) The fuzzy structure-function model. In the fuzzy inference system, the logical relationship is established between the fuzzy sets of the input and output (*top*). In proteins, fuzzification generates sets of interaction patterns amongst functional sequence motifs, which can be linked to conformational sub-states. The connection between structure and function is a knowledge-based logical rule between the set of conformational sub-states and the set of alternative functions, from which the most likely activity can be selected (*bottom*).

6. Conformational Heterogeneity

Dynamic signals perturb conformational ensembles by changing the relative populations of the different sub-states [43] (Figure 1B). The co-existence of functionally different conformations, in a broad regime, may enable the same protein to be simultaneously engaged in multiple pathways [44]. An agonist binding to a β_2-adrenergic receptor, for example, does not stabilize the active conformation of the cytoplasmic domain; it rather increases the conformational heterogeneity of the active, intermediate, and inactive states, for the complex signaling outputs [28].

Intriguing observations indicate that specific biomolecular recognition can also be achieved in heterogeneous conformational ensembles [45–47]. Although the underlying molecular forces are often puzzling [48,49], conformational ambiguities often enable context-dependent responses, via alternative interaction patterns [50,51]. Conformational heterogeneity along the binding trajectory, has recently been concluded to critically influence the structures in a complex, with different partners [52,53]. Structural ambiguities might even be a pre-requisite, for example, for efficient transcription [54] via a fuzzy 'free-for-all' mechanism [55].

Conformational heterogeneity often leads to dynamic interaction profiles, where the functional output (specificity, signal, and polymerization) is controlled by transient contacts [56,57]. Dynamic interactions may also balance between the auto-inhibited and active states [58,59] and can be significantly influenced by post-translational modifications (PTMs) [60,61]. Although the modification pattern inducing the functional response can be defined, its impact on the underlying heterogeneous conformational ensembles often remains unclear.

7. Redundant Sequence Motifs

Multiple, weakly-restrained sequence motifs are frequently distinguished in signaling pathways, via mediating protein interactions [62]. Regions linking the motifs exhibit increased conformational plasticity and reduced sensitivity to mutations or scrambling [63], leading to a phenomenon, often referred to as 'sequence independence' [64]. Tandem repeats of a few residues, for example, are often involved in the organization of higher-order structures [65], ranging from amyloids to signaling complexes and nuclear pores [66]. Motif redundancy leads to the redundancy of interaction patterns and the co-existence of different contact topologies. Although the interactions of the individual motifs are often sub-optimal, their cooperativity may result in high-affinity associations [25,67].

Both the dynamics of the motif-linking regions, and the variations in contact patterns, lead to conformational heterogeneity in higher-order assemblies [68]. The Fused in Sarcoma (Fus) protein, for example, is involved in the formation of stress granules, via a liquid–liquid phase transition, which is driven by its low-complexity (LC) domain, composed of 27 [S/G]Y[S/G] repeats. The NMR spectra of the LC domain in the droplet, is similar to that of the unbound state, witnessing conformational heterogeneity in the assembly [69]. Single-point mutations may gradually decrease conformational heterogeneity, leading to pathological aggregation [70]. Additional studies corroborate the finding that pathological mutations initially induce minor perturbations [71], which simultaneously affect multiple conformations/interaction patterns and induce their shift towards the fibril form.

8. Generalized Structure-Function Ensembles

The experimental data summarized in the above three sections are difficult to interpret via the classical structure-function paradigm (2). We may attempt to solve these problems by treating the sequences, conformations, and functions as generalized ensembles:

$$\text{SEQUENCE}\ (\mu, \sigma) \rightarrow \text{CONFORMATIONAL ENSEMBLE}\ (\mu, \sigma) \rightarrow \text{FUNCTION}\ (\mu, \sigma) \quad (3)$$

where μ is the mean, and σ is the variance of the respective distribution.

Evaluating the structure-function paradigm in the form (3), requires decoupling of all the respective activities, to analyze the underlying distributions of conformational ensembles and sequences. Careful experimental studies, along these lines [72], demonstrate that these approaches are hardly feasible. First, because the dimensionality of the problem is overwhelming, and second, the deconvolution of different functionalities may not be possible in vivo, owing to the intricate connections.

9. Fuzzy Sets Quantify Sequence and Conformation Ambiguities

I propose that the fuzzy set theory [73] offers a quantitative framework to derive stochastic structure-function relationships. In fuzzy sets $U = \{x_1, x_2, \ldots, x_N\}$ a membership function $m(x_i) \rightarrow [0,1]$; $x_i \in U$ is assigned to each element, which characterizes to what extent x_i belongs to the given set. For example, the membership of a protein conformational sub-state (CS_i), in a specific functional set (F_1), can vary between 0 and 1 $(m_1(CS_i) : F_1 \rightarrow [0,1])$, allowing the conformation to contribute to additional activities (e.g., F_2 and F_3, Figure 1C). Memberships for other possible biological functions could also be defined, using this formalism (Figure 1C). In a similar manner, memberships of sequences in given conformational ensembles, $(m_1(SEQ_i) : CS_1 \rightarrow [0,1])$, or in given functions $(m_1(SEQ_i) : F_1 \rightarrow [0,1])$, could also be quantified.

The structure-function paradigm could thus, be reformulated by treating the sequences and conformational ensembles as fuzzy sets:

$$\begin{array}{ll} \text{SEQUENCE} \rightarrow \text{CONFORMATIONAL ENSEMBLE} \rightarrow \text{FUNCTION} & (4) \\ \mathrm{m_i(PI)} : \mathrm{CS_i} \rightarrow [0,1] \qquad \mathrm{m_i(CS)} : \mathrm{F_i} \rightarrow [0,1] & \\ \mathrm{m_i(PI)} : \mathrm{F_i} \rightarrow [0,1] & \end{array}$$

where $m_i(PI)$ is the respective membership function of a sequence, defined with respect to the conformational states (CS_i) or biological activity (F_i), as a pattern of interacting elements/motifs (*PI*). $m_i(CS)$ is the membership function of the conformational sub-state/ensemble (*CS*), in a given function.

Here sequence, structure, and function are considered as different co-existing distributions (Figure 1C), and their contributions change according to the cellular conditions. For example, in the case of a β_2-adrenergic receptor, the active, intermediate, and inactive states (represented by three ensembles) are mixed differently, depending on the signaling input. The fuzzy formalism handles combinations of activities aiming to determine the individual contributions of the different conformational ensembles.

10. The Stochastic Structure-Function Relationship

Within this framework, the structure-function relationship can be quantified by a fuzzy inference system [74,75] (Figure 1D). Parameters describing the elements of the sequence (motifs) or conformational space (distinguished secondary structures) are used as the input, and the different biological activities serve as the output of the system. The first step is the fuzzification of the input, when the fuzzy sets and their membership functions are defined to describe the interaction patterns, and the corresponding conformational sub-states (Figure 1D). The fuzzy inputs are then combined and knowledge-based logical rules ('IF-THEN') are applied to obtain the output membership functions of the different biological activities in the system. These rules could be derived using machine-learning or neural network algorithms. Defuzzification of the output can select the most likely activity, under a given condition, while also accounting for other, promiscuous activities (Figure 1D).

The fuzzy model quantifies the functional ambiguities of the conformational sub-states:

$$\Delta F_{main} = (\sum_{i}^{n} \delta F_{i,main})/n \tag{5}$$

where n is the number of alternative (promiscuous) activities, and F_{main} is the main function with membership function m_{max}. The contribution of function F_i, with respect to the main function, is computed from the corresponding membership functions: $\delta F_{i,\ main} = m_i/\ m_{max}$.

Here, the challenge is to define the membership functions. To this end, the efficiencies of the alternative activities (e.g., catalytic rates) are determined via functional assays on well-characterized conformations (e.g., crystal structures, chip-bound proteins, or those selected by conformational antibodies) or ensembles (solution techniques, NMR, FRET, and single-molecule methods). Different membership functions could be probed computationally, based on the regulatory characteristics (e.g., changing an auto-inhibited to an active state).

The fuzzy formalism (4) is particularly useful to relate the sequence sets to function. Here structural features, which could be predicted from the sequence (e.g., secondary structure elements, disordered regions, or post-translational modification sites) may serve to generate the pattern of interaction elements (PI), to define the fuzzy sets. This approach has been implemented in simulations of higher-order protein organizations [76].

11. Conclusion and Outlook

Proteins deal with uncertain information, regarding cellular conditions. The information is not only imprecise, but various components are unknown or are unpredictable, owing to the non-random fluctuations in the system. The functional characteristics of proteins need to be adjusted to this poorly defined environment. The classical models in protein science, such as the structure-function paradigm, are based on well-defined properties and cannot deal with the ambiguities related to "noise". The fuzzy set theory offers a quantitative framework to reformulate the structure-function paradigm for describing the stochastic cellular behavior of proteins (Figure 1D). This approach will provide a more holistic protein model, which can be applied to generate interaction or metabolic networks of different cell lines as well as more reliable pharmacophore models.

Funding: Financial support is provided by GINOP-2.3.2-15-2016-00044, HAS 11015 and the DE Excellence Program.

Conflicts of Interest: The author declared no conflicts of interest.

References

1. Perutz, M.F. Relation between structure and sequence of haemoglobin. *Nature* **1962**, *194*, 914–917. [CrossRef] [PubMed]
2. Kendrew, J.C. Myoglobin and the structure of proteins. *Science* **1963**, *139*, 1259–1266. [CrossRef] [PubMed]
3. Koshland, D.E., Jr. Properties of the active site of enzymes. *Ann. N Y Acad. Sci.* **1963**, *103*, 630–642. [CrossRef] [PubMed]

4. Blake, C.C.; Koenig, D.F.; Mair, G.A.; North, A.C.; Phillips, D.C.; Sarma, V.R. Structure of hen egg-white lysozyme. A three-dimensional Fourier synthesis at 2 Angstrom resolution. *Nature* **1965**, *206*, 757–761. [CrossRef] [PubMed]
5. Warshel, A. Energetics of enzyme catalysis. *Proc. Natl. Acad. Sci. USA* **1978**, *75*, 5250–5254. [CrossRef] [PubMed]
6. Blomberg, R.; Kries, H.; Pinkas, D.M.; Mittl, P.R.; Grutter, M.G.; Privett, H.K.; Mayo, S.L.; Hilvert, D. Precision is essential for efficient catalysis in an evolved Kemp eliminase. *Nature* **2013**, *503*, 418–421. [CrossRef] [PubMed]
7. Frauenfelder, H.; Sligar, S.G.; Wolynes, P.G. The energy landscapes and motions of proteins. *Science* **1991**, *254*, 1598–1603. [CrossRef] [PubMed]
8. Artymiuk, P.J.; Blake, C.C.; Grace, D.E.; Oatley, S.J.; Phillips, D.C.; Sternberg, M.J. Crystallographic studies of the dynamic properties of lysozyme. *Nature* **1979**, *280*, 563–568. [CrossRef] [PubMed]
9. Frauenfelder, H.; Petsko, G.A.; Tsernoglou, D. Temperature-dependent X-ray diffraction as a probe of protein structural dynamics. *Nature* **1979**, *280*, 558–563. [CrossRef] [PubMed]
10. Henzler-Wildman, K.; Kern, D. Dynamic personalities of proteins. *Nature* **2007**, *450*, 964–972. [CrossRef] [PubMed]
11. Lewandowski, J.R.; Halse, M.E.; Blackledge, M.; Emsley, L. Protein dynamics. Direct observation of hierarchical protein dynamics. *Science* **2015**, *348*, 578–581. [CrossRef] [PubMed]
12. Fontes, M.R.; Teh, T.; Kobe, B. Structural basis of recognition of monopartite and bipartite nuclear localization sequences by mammalian importin-alpha. *J. Mol. Biol.* **2000**, *297*, 1183–1194. [CrossRef] [PubMed]
13. Diez, M.; Zimmermann, B.; Borsch, M.; Konig, M.; Schweinberger, E.; Steigmiller, S.; Reuter, R.; Felekyan, S.; Kudryavtsev, V.; Seidel, C.A.; et al. Proton-powered subunit rotation in single membrane-bound F0F1-ATP synthase. *Nat. Struct. Mol. Biol.* **2004**, *11*, 135–141. [CrossRef] [PubMed]
14. Lindorff-Larsen, K.; Best, R.B.; Depristo, M.A.; Dobson, C.M.; Vendruscolo, M. Simultaneous determination of protein structure and dynamics. *Nature* **2005**, *433*, 128–132. [CrossRef] [PubMed]
15. Jensen, M.R.; Zweckstetter, M.; Huang, J.R.; Blackledge, M. Exploring free-energy landscapes of intrinsically disordered proteins at atomic resolution using NMR spectroscopy. *Chem. Rev.* **2014**, *114*, 6632–6660. [CrossRef] [PubMed]
16. Zhang, J.; Maslov, S.; Shakhnovich, E.I. Constraints imposed by non-functional protein-protein interactions on gene expression and proteome size. *Mol. Syst. Biol.* **2008**, *4*, 210. [CrossRef] [PubMed]
17. Levy, E.D.; De, S.; Teichmann, S.A. Cellular crowding imposes global constraints on the chemistry and evolution of proteomes. *Proc. Natl. Acad. Sci. USA* **2012**, *109*, 20461–20466. [CrossRef] [PubMed]
18. McGuffee, S.R.; Elcock, A.H. Diffusion, crowding & protein stability in a dynamic molecular model of the bacterial cytoplasm. *PLoS Comput. Biol.* **2010**, *6*, e1000694.
19. Tawfik, D.S. Messy biology and the origins of evolutionary innovations. *Nat. Chem. Biol.* **2010**, *6*, 692–696. [CrossRef] [PubMed]
20. Butner, K.A.; Kirschner, M.W. Tau protein binds to microtubules through a flexible array of distributed weak sites. *J. Cell. Biol.* **1991**, *115*, 717–730. [CrossRef] [PubMed]
21. Sigler, P.B. Transcriptional activation. Acid blobs and negative noodles. *Nature* **1988**, *333*, 210–212. [CrossRef] [PubMed]
22. Pricer, R.; Gestwicki, J.E.; Mapp, A.K. From Fuzzy to Function: The New Frontier of Protein-Protein Interactions. *Acc. Chem. Res.* **2017**, *50*, 584–589. [CrossRef] [PubMed]
23. Levy, E.D.; Boeri Erba, E.; Robinson, C.V.; Teichmann, S.A. Assembly reflects evolution of protein complexes. *Nature* **2008**, *453*, 1262–1265. [CrossRef] [PubMed]
24. Wu, H. Higher-order assemblies in a new paradigm of signal transduction. *Cell* **2013**, *153*, 287–292. [CrossRef] [PubMed]
25. Bienz, M. Signalosome assembly by domains undergoing dynamic head-to-tail polymerization. *Trends Biochem. Sci.* **2014**, *39*, 487–495. [CrossRef] [PubMed]
26. Franzmann, T.M.; Jahnel, M.; Pozniakovsky, A.; Mahamid, J.; Holehouse, A.S.; Nuske, E.; Richter, D.; Baumeister, W.; Grill, S.W.; Pappu, R.V.; et al. Phase separation of a yeast prion protein promotes cellular fitness. *Science* **2018**, *359*, 5654. [CrossRef] [PubMed]
27. Copley, S.D. Shining a light on enzyme promiscuity. *Curr. Opin. Struct. Biol.* **2017**, *47*, 167–175. [CrossRef] [PubMed]

28. Manglik, A.; Kim, T.H.; Masureel, M.; Altenbach, C.; Yang, Z.; Hilger, D.; Lerch, M.T.; Kobilka, T.S.; Thian, F.S.; Hubbell, W.L.; et al. Structural Insights into the Dynamic Process of β_2-Adrenergic Receptor Signaling. *Cell* **2015**, *161*, 1101–1111. [CrossRef] [PubMed]
29. Ross, E.D.; Baxa, U.; Wickner, R.B. Scrambled prion domains form prions and amyloid. *Mol. Cell. Biol.* **2004**, *24*, 7206–7213. [CrossRef] [PubMed]
30. Khersonsky, O.; Tawfik, D.S. Enzyme promiscuity: A mechanistic and evolutionary perspective. *Annu. Rev. Biochem.* **2010**, *79*, 471–505. [PubMed]
31. Jeffery, C.J. Moonlighting proteins: Old proteins learning new tricks. *Trends Genet.* **2003**, *19*, 415–417. [CrossRef]
32. Freilich, S.; Spriggs, R.V.; George, R.A.; Al-Lazikani, B.; Swindells, M.; Thornton, J.M. The complement of enzymatic sets in different species. *J. Mol. Biol.* **2005**, *349*, 745–763. [CrossRef] [PubMed]
33. Nam, H.; Lewis, N.E.; Lerman, J.A.; Lee, D.H.; Chang, R.L.; Kim, D.; Palsson, B.O. Network context and selection in the evolution to enzyme specificity. *Science* **2012**, *337*, 1101–1104. [CrossRef] [PubMed]
34. Arnold, F.H.; Wintrode, P.L.; Miyazaki, K.; Gershenson, A. How enzymes adapt: Lessons from directed evolution. *Trends Biochem. Sci.* **2001**, *26*, 100–106. [CrossRef]
35. Tokuriki, N.; Jackson, C.J.; Afriat-Jurnou, L.; Wyganowski, K.T.; Tang, R.; Tawfik, D.S. Diminishing returns and tradeoffs constrain the laboratory optimization of an enzyme. *Nat. Commun.* **2012**, *3*, 1257. [CrossRef] [PubMed]
36. Aharoni, A.; Gaidukov, L.; Khersonsky, O.; Mc, Q.G.S.; Roodveldt, C.; Tawfik, D.S. The 'evolvability' of promiscuous protein functions. *Nat. Genet.* **2005**, *37*, 73–76. [CrossRef] [PubMed]
37. Amitai, G.; Gupta, R.D.; Tawfik, D.S. Latent evolutionary potentials under the neutral mutational drift of an enzyme. *HFSP J.* **2007**, *1*, 67–78. [CrossRef] [PubMed]
38. James, L.C.; Roversi, P.; Tawfik, D.S. Antibody multispecificity mediated by conformational diversity. *Science* **2003**, *299*, 1362–1367. [CrossRef] [PubMed]
39. Campbell, E.; Kaltenbach, M.; Correy, G.J.; Carr, P.D.; Porebski, B.T.; Livingstone, E.K.; Afriat-Jurnou, L.; Buckle, A.M.; Weik, M.; Hollfelder, F.; et al. The role of protein dynamics in the evolution of new enzyme function. *Nat. Chem. Biol.* **2016**, *12*, 944–950. [CrossRef] [PubMed]
40. Brooks, B.; Karplus, M. Harmonic dynamics of proteins: Normal modes and fluctuations in bovine pancreatic trypsin inhibitor. *Proc. Natl. Acad. Sci. USA* **1983**, *80*, 6571–6575. [CrossRef] [PubMed]
41. Marsh, J.A.; Teichmann, S.A. Parallel dynamics and evolution: Protein conformational fluctuations and assembly reflect evolutionary changes in sequence and structure. *Bioessays* **2014**, *36*, 209–218. [CrossRef] [PubMed]
42. Haliloglu, T.; Bahar, I. Adaptability of protein structures to enable functional interactions and evolutionary implications. *Curr. Opin. Struct. Biol.* **2015**, *35*, 17–23. [CrossRef] [PubMed]
43. Smock, R.G.; Gierasch, L.M. Sending signals dynamically. *Science* **2009**, *324*, 198–203. [CrossRef] [PubMed]
44. Nygaard, R.; Zou, Y.; Dror, R.O.; Mildorf, T.J.; Arlow, D.H.; Manglik, A.; Pan, A.C.; Liu, C.W.; Fung, J.J.; Bokoch, M.P.; et al. The dynamic process of beta(2)-adrenergic receptor activation. *Cell* **2013**, *152*, 532–542. [CrossRef] [PubMed]
45. Tompa, P.; Fuxreiter, M. Fuzzy complexes: Polymorphism and structural disorder in protein-protein interactions. *Trends Biochem. Sci.* **2008**, *33*, 2–8. [CrossRef] [PubMed]
46. Sharma, R.; Raduly, Z.; Miskei, M.; Fuxreiter, M. Fuzzy complexes: Specific binding without complete folding. *FEBS Lett.* **2015**, *589*, 2533–2542. [CrossRef] [PubMed]
47. Fuxreiter, M. Fuzziness in Protein Interactions-A Historical Perspective. *J. Mol. Biol.* **2018**, *430*, 2278–2287. [CrossRef] [PubMed]
48. Hadzi, S.; Mernik, A.; Podlipnik, C.; Loris, R.; Lah, J. The Thermodynamic Basis of the Fuzzy Interaction of an Intrinsically Disordered Protein. *Angew Chem. Int. Ed. Engl.* **2017**, *56*, 14494–14497. [CrossRef] [PubMed]
49. Fuxreiter, M.; Simon, I.; Bondos, S. Dynamic protein-DNA recognition: Beyond what can be seen. *Trends Biochem. Sci.* **2011**, *36*, 415–423. [CrossRef] [PubMed]
50. Buljan, M.; Chalancon, G.; Eustermann, S.; Wagner, G.P.; Fuxreiter, M.; Bateman, A.; Babu, M.M. Tissue-Specific Splicing of Disordered Segments that Embed Binding Motifs Rewires Protein Interaction Networks. *Mol. Cell* **2012**, *46*, 871–883. [CrossRef] [PubMed]
51. Miskei, M.; Gregus, A.; Sharma, R.; Duro, N.; Zsolyomi, F.; Fuxreiter, M. Fuzziness enables context dependence of protein interactions. *FEBS Lett.* **2017**, *591*, 2682–2695. [CrossRef] [PubMed]

52. Toto, A.; Camilloni, C.; Giri, R.; Brunori, M.; Vendruscolo, M.; Gianni, S. Molecular Recognition by Templated Folding of an Intrinsically Disordered Protein. *Sci. Rep.* **2016**, *6*, 21994. [CrossRef] [PubMed]
53. Fuxreiter, M. Fold or not to fold upon binding-does it really matter? *Curr. Opin. Struct. Biol.* **2018**, *54*, 19–25. [CrossRef] [PubMed]
54. Warfield, L.; Tuttle, L.M.; Pacheco, D.; Klevit, R.E.; Hahn, S. A sequence-specific transcription activator motif and powerful synthetic variants that bind Mediator using a fuzzy protein interface. *Proc. Natl. Acad. Sci. USA* **2014**, *111*, E3506–E3513. [CrossRef] [PubMed]
55. Tuttle, L.M.; Pacheco, D.; Warfield, L.; Luo, J.; Ranish, J.; Hahn, S.; Klevit, R.E. Gcn4-Mediator Specificity Is Mediated by a Large and Dynamic Fuzzy Protein-Protein Complex. *Cell Rep.* **2018**, *22*, 3251–3264. [CrossRef] [PubMed]
56. Delaforge, E.; Kragelj, J.; Tengo, L.; Palencia, A.; Milles, S.; Bouvignies, G.; Salvi, N.; Blackledge, M.; Jensen, M.R. Deciphering the Dynamic Interaction Profile of an Intrinsically Disordered Protein by NMR Exchange Spectroscopy. *J. Am. Chem. Soc.* **2018**, *140*, 1148–1158. [CrossRef] [PubMed]
57. Kragelj, J.; Palencia, A.; Nanao, M.H.; Maurin, D.; Bouvignies, G.; Blackledge, M.; Jensen, M.R. Structure and dynamics of the MKK7-JNK signaling complex. *Proc. Natl. Acad. Sci. USA.* **2015**, *112*, 3409–3414. [CrossRef] [PubMed]
58. Gogl, G.; Alexa, A.; Kiss, B.; Katona, G.; Kovacs, M.; Bodor, A.; Remenyi, A.; Nyitray, L. Structural Basis of Ribosomal S6 Kinase 1 (RSK1) Inhibition by S100B Protein: Modulation of the extracellular signal-regulated kinase (ERK) signaling cascade in a calcium-dependent way. *J. Biol. Chem.* **2016**, *291*, 11–27. [CrossRef] [PubMed]
59. Lu, A.; Magupalli, V.G.; Ruan, J.; Yin, Q.; Atianand, M.K.; Vos, M.R.; Schroder, G.F.; Fitzgerald, K.A.; Wu, H.; Egelman, E.H. Unified polymerization mechanism for the assembly of ASC-dependent inflammasomes. *Cell* **2014**, *156*, 1193–1206. [CrossRef] [PubMed]
60. Lee, G.M.; Pufall, M.A.; Meeker, C.A.; Kang, H.S.; Graves, B.J.; McIntosh, L.P. The affinity of Ets-1 for DNA is modulated by phosphorylation through transient interactions of an unstructured region. *J. Mol. Biol.* **2008**, *382*, 1014–1030. [CrossRef] [PubMed]
61. Li, J.; McQuade, T.; Siemer, A.B.; Napetschnig, J.; Moriwaki, K.; Hsiao, Y.S.; Damko, E.; Moquin, D.; Walz, T.; McDermott, A.; et al. The RIP1/RIP3 Necrosome Forms a Functional Amyloid Signaling Complex Required for Programmed Necrosis. *Cell* **2012**, *150*, 339–350. [CrossRef] [PubMed]
62. Davey, N.E.; Cyert, M.S.; Moses, A.M. Short linear motifs-ex nihilo evolution of protein regulation. *Cell Commun. Signal.* **2015**, *13*, 43. [CrossRef] [PubMed]
63. Lu, X.; Hamkalo, B.; Parseghian, M.H.; Hansen, J.C. Chromatin condensing functions of the linker histone C-terminal domain are mediated by specific amino acid composition and intrinsic protein disorder. *Biochemistry* **2009**, *48*, 164–172. [CrossRef] [PubMed]
64. Ross, E.D.; Edskes, H.K.; Terry, M.J.; Wickner, R.B. Primary sequence independence for prion formation. *Proc. Natl. Acad. Sci. USA* **2005**, *102*, 12825–12830. [CrossRef] [PubMed]
65. Nott, T.J.; Petsalaki, E.; Farber, P.; Jervis, D.; Fussner, E.; Plochowietz, A.; Craggs, T.D.; Bazett-Jones, D.P.; Pawson, T.; Forman-Kay, J.D.; et al. Phase transition of a disordered nuage protein generates environmentally responsive membraneless organelles. *Mol. Cell* **2015**, *57*, 936–947. [CrossRef] [PubMed]
66. Milles, S.; Mercadante, D.; Aramburu, I.V.; Jensen, M.R.; Banterle, N.; Koehler, C.; Tyagi, S.; Clarke, J.; Shammas, S.L.; Blackledge, M.; et al. Plasticity of an Ultrafast Interaction between Nucleoporins and Nuclear Transport Receptors. *Cell* **2015**, *163*, 734–745. [CrossRef] [PubMed]
67. Ciuffa, R.; Lamark, T.; Tarafder, A.K.; Guesdon, A.; Rybina, S.; Hagen, W.J.; Johansen, T.; Sachse, C. The selective autophagy receptor p62 forms a flexible filamentous helical scaffold. *Cell Rep.* **2015**, *11*, 748–758. [CrossRef] [PubMed]
68. Wu, H.; Fuxreiter, M. The Structure and Dynamics of Higher-Order Assemblies: Amyloids, Signalosomes, and Granules. *Cell* **2016**, *165*, 1055–1066. [CrossRef] [PubMed]
69. Burke, K.A.; Janke, A.M.; Rhine, C.L.; Fawzi, N.L. Residue-by-Residue View of In Vitro FUS Granules that Bind the C-Terminal Domain of RNA Polymerase II. *Mol. Cell* **2015**, *60*, 231–241. [CrossRef] [PubMed]
70. Patel, A.; Lee, H.O.; Jawerth, L.; Maharana, S.; Jahnel, M.; Hein, M.Y.; Stoynov, S.; Mahamid, J.; Saha, S.; Franzmann, T.M.; et al. A Liquid-to-Solid Phase Transition of the ALS Protein FUS Accelerated by Disease Mutation. *Cell* **2015**, *162*, 1066–1077. [CrossRef] [PubMed]

71. Molliex, A.; Temirov, J.; Lee, J.; Coughlin, M.; Kanagaraj, A.P.; Kim, H.J.; Mittag, T.; Taylor, J.P. Phase Separation by Low Complexity Domains Promotes Stress Granule Assembly and Drives Pathological Fibrillization. *Cell* **2015**, *163*, 123–133. [CrossRef] [PubMed]
72. Nagulapalli, M.; Parigi, G.; Yuan, J.; Gsponer, J.; Deraos, G.; Bamm, V.V.; Harauz, G.; Matsoukas, J.; de Planque, M.R.; Gerothanassis, I.P.; et al. Recognition pliability is coupled to structural heterogeneity: A calmodulin intrinsically disordered binding region complex. *Structure* **2012**, *20*, 522–533. [CrossRef] [PubMed]
73. Zadeh, L.A. Fuzzy sets. *Inf. Control* **1965**, *8*, 338–353. [CrossRef]
74. Zadeh, L.A. Outline of a new approach to the analysis of complex systems and decision processes. *IEEE Trans. Syst. Man Cybern.* **1973**, *3*, 28–44. [CrossRef]
75. Mamdani, E.H.; Assilian, S. An experiment in linguistic synthesis with a fuzzy logic controller. *Int. J. Man-Mach. Stud.* **1975**, *7*, 1–13. [CrossRef]
76. Tuu-Szabo, B.; Koczy, L.; Fuxreiter, M. Simulations of Higher-Order Protein Organizations Using a Fuzzy Framework. *Complexity* **2018**, in press.

molecules

MDPI

Review

Moonlighting Proteins in the Fuzzy Logic of Cellular Metabolism

Haipeng Liu [1] and Constance J. Jeffery [2,*]

1 Center for Biomolecular Sciences, College of Pharmacy, University of Illinois at Chicago, 900 South Ashland Avenue, Chicago, IL 60607, USA; hliu47@uic.edu

2 Department of Biological Sciences, University of Illinois at Chicago, 900 South Ashland Avenue, Chicago, IL 60607, USA

* Correspondence: cjeffery@uic.edu; Tel.: +1-312-996-3168

Academic Editor: Pier Luigi Gentili

Received: 30 May 2020; Accepted: 23 July 2020; Published: 29 July 2020

Abstract: The numerous interconnected biochemical pathways that make up the metabolism of a living cell comprise a fuzzy logic system because of its high level of complexity and our inability to fully understand, predict, and model the many activities, how they interact, and their regulation. Each cell contains thousands of proteins with changing levels of expression, levels of activity, and patterns of interactions. Adding more layers of complexity is the number of proteins that have multiple functions. Moonlighting proteins include a wide variety of proteins where two or more functions are performed by one polypeptide chain. In this article, we discuss examples of proteins with variable functions that contribute to the fuzziness of cellular metabolism.

Keywords: moonlighting proteins; fuzzy logic; intrinsically disordered proteins; metamorphic proteins; morpheeins

1. Introduction

Fuzzy logic systems include variables that can be any real number between 0 and 1 instead of being limited to the Boolean logic variables of only 0 and 1. This enables expression of complexity, uncertainty, and imprecision. In general, the vast interconnected biochemical pathways that make up the metabolism of a living cell can appear fuzzy because they are complex and hard to predict, and we have incomplete and not yet accurate knowledge. A single cell contains thousands of proteins performing a wide variety of activities, and the proteins have complex and constantly changing levels of expression, levels of activity, and patterns of interactions with other proteins and other molecules. Adding even more layers of complexity is the ability of many proteins, called moonlighting proteins, to perform more than one function. Moonlighting proteins are proteins in which one polypeptide chain performs more than one physiologically relevant biochemical or biophysical function [1–3] (Figure 1). The MoonProt Database (www.moonlightingproteins.org) contains annotations for over 300 experimentally confirmed moonlighting proteins, of which about 130 proteins are from human [4,5]. Although the mechanisms by which one protein performs two different functions are not always understood, it is clear that the function (or functions) performed at any specific time can be affected by multiple factors, and sometimes combinations of factors, including targeting to different cellular compartments, changes in the intracellular concentration of ligands, and changes in environmental conditions. In this paper, we describe examples of moonlighting proteins and some of the mechanisms by which they change function. These examples help illustrate and complement the ideas in this collection of papers on the topic of "The Fuzziness in Molecular Supramolecular, and Systems Chemistry", where Gentili presents the "Fuzziness of the Molecular World" and describes natural information systems that involve fuzzy logic in large part due to proteins having multiple features and functions that vary in a context-dependent manner [6]. In addition, the paper by Fuxreiter describes using fuzzy set theory in a quantitative framework for describing the relationships between changing protein structures, interactions, and functions under changing, and somewhat unknown or unpredictable, cellular conditions [7].

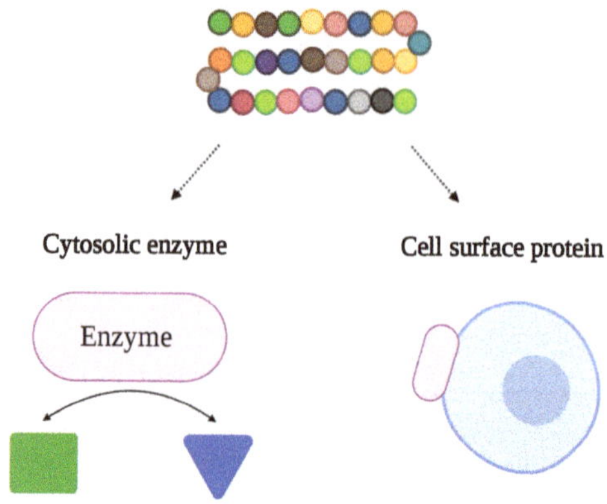

Figure 1. In a moonlighting protein (purple oval), more than one physiologically relevant biochemical or biophysical function is performed by a single polypeptide chain. Note: This figure was "Created with BioRender.com".

2. Examples of Moonlighting Proteins and Factors that Affect Function

2.1. Cellular Localization

The most often observed subclass of moonlighting protein includes proteins that perform different functions in different cellular localizations. Over 100 enzymes and chaperones that catalyze reactions in the cytosol can be secreted and act as cytokines that modify the host's immune system or become bound to the cell membrane where they serve as cell surface receptors, and in some cases these second functions contribute to virulence [8–11].

Enolase is one of these intracellular/surface moonlighting proteins in many species, including eukaryotes as well as prokaryotes. Inside the cell, it catalyzes the conversion of 2-phosphoglycerate to phosphoenolpyruvate in glycolysis. When displayed on the cell surface, it binds to host proteins (Figure 2a). The enolases from *Aeromonas hydrophila*, *Bacillus anthracis*, *Neisseria meningitidis*, *Streptococcus pneumoniae*, *Trichomoniasis vaginalis* and *Lactobacillus crispatus* can bind to host plasminogen [12–17]. The binding of plasminogen plays an important role in invasion of host tissues because, once bound to the cell surface receptor, the plasminogen becomes converted to the active protease, plasmin, which can aid in breaking down host extracellular matrix and invasion of tissues [18,19]. In some species, surface-located enolase and other intracellular/surface moonlighting proteins bind to other host proteins for colonization or for modulating the host immune system. *Streptococcus suis* enolase can also bind to host fibronectin, and *Staphylococcus aureus* enolase exhibits laminin binding activity [20,21].

Glyceraldehyde 3-phosphate dehydrogenase (GAPDH) is another commonly found intracellular/surface moonlighting protein. It catalyzes the conversion of glyceraldehyde 3-phosphate to glycerate 1,3-bisphosphate in glycolysis in the cytoplasm. Some commensal bacteria that colonize the human gut use GAPDH on the cell surface to bind to host mucin and enable colonization of the gut [22,23]. When expressed on the cell surface, *Streptococcus pyogenes* GADPH can bind to plasminogen and fibronectin, and can also function as a ADP-ribosylating enzyme and assist with neutrophil evasion [24–27]. In addition, *Streptococcus agalactiae* GADPH can act as a modulator of the host's immune system [28].

As the bacterial HSP70 (heat shock protein 70), DnaK is abundantly expressed in the cytosol as a stress-inducible chaperone. *Mycobacterium tuberculosis* DnaK can be displayed on the cell surface to bind to plasminogen, and it can also bind to CD40 (cluster of differentiation 40) to stimulate the synthesis of monocyte chemokines and the maturation of dendritic cells [29,30].

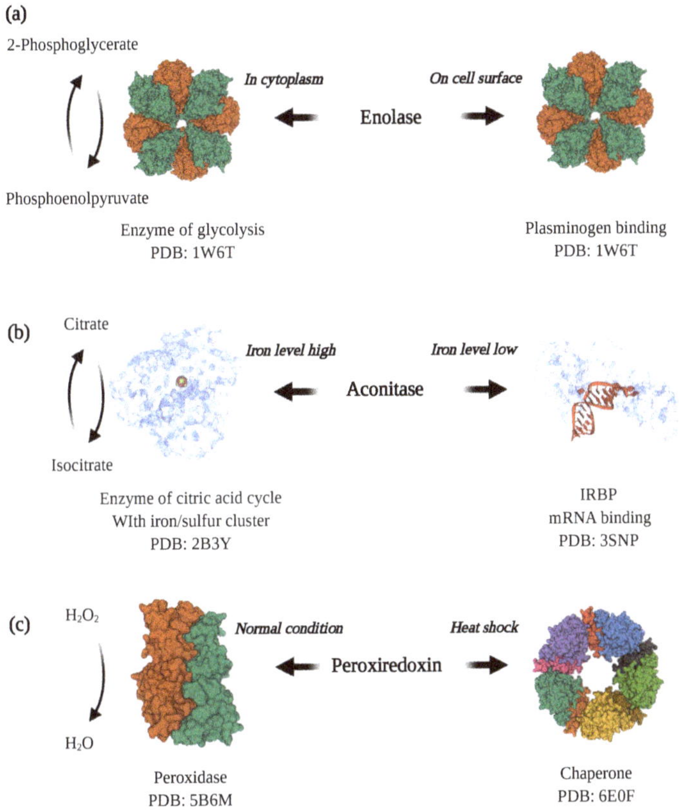

Figure 2. Examples of moonlighting proteins. (**a**) Enolase is a cytosolic enzyme in glycolysis and also a plasminogen receptor when displayed on the cell surface (PDB ID: 1W6T [31]). (**b**) Aconitase is an enzyme in the citric acid cycle when it contains an iron/sulfur cluster bound in the active site of the protein (PDB ID: 2B3Y [32]). When the cellular iron level decreases and the iron/sulfur cluster disassociates, aconitase undergoes a large conformational change that enables it to bind to iron-responsive elements in mRNA (PDB ID: 3SNP [33]) to promote the expression of proteins involved in iron uptake. (**c**) Under normal cellular conditions, peroxiredoxin is predominantly a dimer (PDB ID: 5B6M [34]) that functions as a peroxidase that converts hydrogen peroxide to water. Under heat shock or oxidative stress, it converts to a higher molecular weight form, a decamer, that acts as a molecular chaperone that assists with protein folding (PDB ID: 6E0F [35]). Note: This figure was "Created with BioRender.com", and the visualizations of the protein structures were created with Mol* [36] on the RCSB PDB website (rcsb.org) [37,38].

2.2. *Interactions with Other Proteins and Molecules*

Changes in cellular concentration of substrates or other ligands can serve as a trigger for changing protein functions. Aconitase is an enzyme in the citric acid cycle that contains an iron–sulfur cluster in the active site (Figure 2b). When the intracellular concentration of iron is high, the iron–sulfur cluster in the active site enables the isomerization of citrate to isocitrate [39]. In contrast, when the cellular level of iron is low, the iron–sulfur cluster is lost, and the enzyme changes conformation to expose an RNA binding surface and becomes an iron-responsive element binding protein (IRBP). As an IRBP, it binds to Iron Responsive Element (IRE) sequence motifs in mRNA, leading to increased translation of proteins involved in cellular iron uptake [40,41]. Another example is BirA, which performs functions as an enzyme and a transcription repressor in the biotin regulatory system [42]. The enzyme's function is determined by the cellular need for biotin. When the need is high, when the cells are growing rapidly, BirA functions as a biotin–[acetyl–CoA-carboxylase] ligase that transfers a biotinyl moiety to the biotin carboxyl carrier protein subunit of acetyl–CoA carboxylase. When the demand decreases, under slower growth rates, BirA binds to DNA and functions as a biotin–operon repressor to inhibit the production of biotin [43].

As with aconitase and BirA, the different functions of many moonlighting proteins require interactions with different proteins, multiprotein complexes, DNA, RNA and other macromolecules. Many ribosomal proteins are moonlighting proteins, such as ribosomal proteins S3, S13, S14, L2, L4, L5, L7, L11, L13a, L23 and L26 [44–54]. These proteins have a second function when they dissociate from the ribosome and interact with other molecules. For example, the ribosomal protein S3 is a component of the 40S subunit of the ribosome, which is located in the cytoplasm. When S3 dissociates from the ribosome, it can enter the nucleus and act as a deoxyribonuclease (DNase) that cleaves apurinic/apyrimidinic sites during DNA repair [55]. Like the cytosolic aconitase mentioned above, mitochondrial aconitase acts as an enzyme in the citric acid cycle to catalyze the isomerization of citrate to isocitrate. However, instead of a function in translation, mitochondrial aconitase plays an important role in the maintenance of mitochondrial DNA that is independent of its catalytic role in the citric acid cycle [56].

2.3. *Environmental Stress*

Another common factor for moonlighting proteins to switch functions is environmental stresses. One example is peroxiredoxin, which in many species changes function from an enzyme to a protein-folding chaperone in response to oxidative stress or heat shock [57] (Figure 2c). In a lower molecular weight form, peroxiredoxin acts as a peroxidase that reduces hydrogen peroxide to water. However, under stress conditions, peroxiredoxin undergoes a shift to a higher molecular weight homo-oligomeric complex, comprised of five dimers connected by hydrophobic interactions [35]. The high molecular weight complex is a molecular chaperone that helps with folding and stabilizing proteins disrupted by the cell stress conditions.

Another example is the DegP protease, which also has a temperature-dependent change of function, where it transitions between a protease and a molecular chaperone. Under low temperature conditions, DegP functions as a molecular chaperone with the proteolytic site inactivated. When temperatures increase, the proteolytic site is activated and DegP can catalyze protein degradation [58].

2.4. *Changes in Protein Structure*

Changes in cellular localization, interaction partners, and environmental conditions can trigger a change in protein function with little or no change in protein structure. For example, in some moonlighting enzymes that are also transcription factors, ligand binding can turn on the transcription factor function by causing a relatively small change in overall structure that increases its DNA binding affinity [59]. Other moonlighting proteins undergo a large conformational change, such as cellular aconitase becoming IRBP, which involves a large relative movement of several domains within a protein subunit to uncover a previously buried mRNA binding surface. The peroxiredoxins described above undergo changes in quaternary structure to become chaperones.

In general, many moonlighting proteins undergo changes in structure, which can range from small movements of surface loops to the more drastic change in tertiary or quaternary structure observed in intrinsically disordered proteins (IDPs), metamorphic proteins, or morpheeins.

Intrinsically disordered proteins have a region or subunit that is unfolded and can in some cases enable a switch from unfolded to multiple folded structures that enable interactions with different proteins. Metamorphic proteins contain a domain or subunit that folds into more than one stable structure. Morpheein proteins have subunits that disassemble, change conformation, and reassemble into different quaternary structures. In some cases, these structural changes are involved in regulation of a single function, but in other cases, the structural changes are correlated with a switch between two different functions.

In this next section, we describe examples of IDPs, metamorphic proteins, and morpheeins and how variability in protein structure contributes to some of these proteins being moonlighting proteins with "fuzziness" in protein function.

2.4.1. Intrinsically Disordered Proteins

Intrinsically disordered proteins contain regions, sometimes the entire polypeptide chain, that are unfolded under physiological conditions. Some unfolded regions are fully functional without becoming completely folded, and others undergo reversible folding when binding with other molecules [60,61]. In some cases, this flexibility enables one IDP to bind to a variety of cellular components, including small molecules, other proteins, DNA, or RNA, which enables many proteins to perform more than one function and contributes to the complexity in cellular metabolism and regulation of transcription and translation, molecular translocation, DNA repair and replication, and cell signaling [62–65].

Thirteen proteins in the database of moonlighting proteins (MoonProt, moonlightingproteins.org) are also found in the database of intrinsically disordered proteins (DisProt.org) (Table 1) [5,66]. Two proteins are discussed further below, p53 and thymosin beta-4.

The tumor suppressor protein p53 is a moonlighting protein and also an IDP with roles in cell cycle regulation, DNA repair and apoptosis [67,68]. In normal cells, the levels of p53 are low, but if cells sense environmental dangers that cause DNA damage, such as toxins, viruses or radiation, the level of p53 rises. Then, p53 binds to regulatory elements in the genome, activating a cascade of cellular responses to stop cell division and prevent cells from uncontrolled growth. In DNA repair, p53 can interact directly with DNA polymerase and AP endonuclease to stimulate base excision repair [69]. Outside the nucleus, p53 also has several cytoplasmic functions, including in centrosome duplication, induction of apoptosis and inhibition of autophagy [68].

In the native state, p53 has both folded and unfolded domains. The folded core domain is a DNA binding domain that recognizes specific regulatory elements. The folded tetramerization domain at the center of the protein joins protein subunits together into a homo-tetramer [70]. The *N*-terminal transactivation domain that interacts with and activates transcription factors is intrinsically disordered [71]. The *C*-terminal domain and the linker regions in between domains are also intrinsically disordered and fairly flexible, which enables the protein to adjust its conformation upon binding to specific regulatory sites in the DNA [72]. The flexibility of the intrinsically disordered domains allows p53 to recognize and bind to a large number of regulatory elements, so that it can regulate transcription in many different sites of the genome.

Thymosin beta-4 (Tβ4) is another moonlighting protein that is also an IDP. Tβ4 mainly functions in sequestering G-actin (monomeric actin) to prevent it from polymerization [73,74]. Tβ4 also has multiple moonlighting functions that are involved in diverse cellular roles including enhancement of endothelial cell differentiation, stimulation of angiogenesis, tissue regeneration and inhibition of inflammatory responses [75–78].

The free form of Tβ4 is intrinsically disordered and predominantly unstructured in solution. However, upon binding with G-actin, Tβ4 becomes fully folded and structured, where an extended conformation in the central region and two helices at the *N*-, *C*-termini can be identified [79,80]. In addition, Tβ4 forms complexes with PINCH (Particularly Interesting New Cys His-containing protein), ILK (Integrin-Linked Kinase) and stabilin-2 (an endocytic receptor for hyaluronic acid) respectively, where weak, transient and structurally ambiguous protein–protein interactions take place [81–83].

Table 1. Examples of moonlighting proteins that contain intrinsically disordered regions.

Protein	Organism	Function 1	Function 1 Location	Function 2	Function 2 Location	Disorder Content	UniProt ID	Ref.
P53	*Homo sapiens*	Binds to regulatory element in the genome	Nucleus	Centrosome duplication, induction of apoptosis, autophagy inhibition	Cytoplasm	37.00%	P04637	[67,69,71]
Thymosin beta-4	*Homo sapiens*	Involved in sequestering G-actin in human polymorphonuclear leukocytes	Cytoplasm	Secreted anti-inflammatory agent	Extracellular	100.00%	P62328	[73,78,84]
General control protein GCN4	*Saccharomyces cerevisiae*	Transcription factor	Nucleus	Ribonuclease	Cytoplasm	32.38%	P03069	[85–87]
Bifunctional ligase/repressor BirA	*Escherichia coli*	Biotin synthetase, biotin–[acetyl–CoA- carboxylase] ligase	Cytoplasm	Biotin operon repressor, activity depends on cellular concentration of biotin	Bound to DNA	7.17%	P06709	[42,43,88, 89]
High mobility group protein B1	*Rattus norvegicus*	Binds heparin.	Cytoplasm	DNA binding protein, without sequence specificity	Nucleus	35.81%	P63159	[90–92]
Calreticulin	*Homo sapiens*	Protein-folding chaperone	Endoplasmic reticulum	Adhesin	Cell surface	100.00%	P27797	[93–95]
50S ribosomal protein L4	*Escherichia coli*	Ribosomal protein, part of the 50S subunit	Cytoplasm	Transcriptional repressor, causes premature termination of transcription	Bound to DNA	31.34%	P60723	[48,96,97]
DNA replication factor Cdt1	*Homo sapiens*	Helps with initiating DNA replication	Nucleus	Role in mitosis, localizes to kinetochores through binding to Ndc80 complex	Cytoplasm	2.20%	Q9H211	[98–100]
Transcriptional regulator Ure2	*Saccharomyce cerevisiae*	Binds to and inhibits GATA transcriptional activators GLN3 and GAT1	Cytoplasm	Glutathione peroxidase, thiol:disulfide oxidoreductase	Cytoplasm	31.64%	P23202	[101–103]
Thymidylate synthase	*Homo sapiens*	Thymidylate synthase	Cytoplasm	mRNA translation inhibition	Cytoplasm	15.65%	P04818	[104–106]
Cytochrome c	*Equus caballus*	Electron carrier protein	Mitochondrion	Binds to apoptosis protease activation factor-1 and promotes apoptosis	Cytoplasm	99.05%	P00004	[107–109]
Band 3 anion transport protein	*Homo sapiens*	Transports inorganic anions across the plasma membrane	Plasma membrane	Scaffold protein providing binding sites for glycolytic enzymes	Plasma membrane	5.93%	P02730	[110–112]
Cystic fibrosis transmembrane conductance regulator CFTR	*Homo sapiens*	Chloride transporter, contains nucleotide binding domains that bind and hydrolyze ATP	Plasma membrane	Regulator of other ion channels	Plasma membrane	12.50%	P13569	[113–115]

Protein structures in which significant conformational heterogeneity, disorder or ambiguity remain after formation of the complex are referred to as "fuzzy interactions" or "fuzzy complexes" and the remaining flexibility or disorder can be important in the assembly or activity of the complexes [116–118]. Fuzzy interactions and complexes can enable interactions with alternative partners and sensitivity to post-translational modifications. Fuzzy interactions also play a large part in several types of supramolecular interactions including in intracellular lipid droplets, which are described in another paper in this collection by Uversky [119].

The fuzzy complexes formed between IDPs and their binding partners, as mentioned above, can be an important feature of IDPs in fulfilling their functional versatility. Moonlighting proteins GCN4, HMGB1, CFTR, and Ure2 are found to be part of fuzzy complexes in the Fuzzy Complex Database (http://protdyn-database.org) [120]. As an example, GCN4 is a transcription activator for several genes, and is also a ribonuclease [86]. As a transcription activator, GCN4 binds Gal11 (an activator) in a weak and low affinity mode with multiple conformations. This conformational ambiguity is a typical example of a fuzzy complex where no single binding conformational state has been identified [117].

2.4.2. Metamorphic Proteins

Metamorphic proteins add another layer of complexity to our understanding of protein structure and function [121,122]. In stark contrast to the dogma of one sequence, one structure, one function, metamorphic proteins have two or more folded structures as their native structures, and in some cases the different structures have different functions. Distinct from the intrinsically disordered proteins, where the native states are folded or unfolded, the native states of metamorphic proteins are both folded and structured. The interconversions between native structures are reversible, meaning that at equilibrium there is a balance between the native structures [123]. Although the fact of two native structures appears to be against the thermodynamic principles of protein folding, in which the native structure of a protein has the overall lowest free energy, it has been shown that the two native structures can have similar energies with a low activation barrier of refolding [124]. So far, a small number of metamorphic proteins have been discovered, including lymphotactin, RfaH, CLIC1, Mad2, KaiB, IscU, Selecase and HIV-1 reverse transcriptase, of which the first three are also moonlighting proteins [125–132].

The C family chemokine lymphotactin (Ltn) is a metamorphic protein and also a moonlighting protein with heparin-binding activity [133]. Under normal physiological conditions, 37 °C and 150 mM NaCl, lymphotactin exists in an equilibrium between two native states, Ltn10 and Ltn40. Ltn10 is a monomer possessing a mix of beta sheet and alpha helix in a canonical chemokine fold that undergoes refolding and dimerization to become Ltn40, which contains a beta sandwich (Figure 3). While Ltn10 is an agonist for the X-C G-protein coupled chemokine receptor 1 (XCR1), Ltn40 can't bind to XCR1 but instead can bind to heparin, a glycosaminoglycan component of the extracellular matrix. At equilibrium, there are nearly equal amounts of Ltn10 and Ltn40. The interconversions between Ltn10 and Ltn40 can be controlled by small changes in salt concentration and temperature. When the salt concentration is high and the temperature is low, the presence of Ltn10 is predominant, however, at lower salt concentrations and higher temperatures, but still below 40 °C, Ltn40 is the predominant species [125]. Most other chemokines don't appear to undergo these transformations because they contain two disulfide bonds. Because lymphotactin only has one disulfide bond, it is less restricted and more flexible in changing conformations compared to other chemokines, which partially explains the reversibility between two distinct native structures. In experiments where an extra disulfide bond was introduced, lymphotactin could be locked in only one native state and the transition to the other state was prohibited, suggesting that a single amino acid modification can change the functionality of lymphotactin significantly [134].

Another metamorphic protein with moonlighting activity is RfaH, which functions as a transcription factor that inhibits termination and is also a translation factor. RfaH has two domains: a C-terminal domain (CTD) and an *N*-terminal domain (NTD). As a transcription factor, these two domains are tightly bound together, and the CTD is in an all-helical conformation that masks the RNAP (RNA polymerase) binding surface on the NTD, preventing the NTD from interacting with RNAP. When the NTD binds to specific operons in DNA, the two domains are separated, which enables the NTD to interact with RNAP. Meanwhile, the CTD undergoes a transformation from an all-helical conformation to an all-beta one, after which the CTD is able to recruit ribosomes and potentiates the translation of operons controlled by RfaH [126,135].

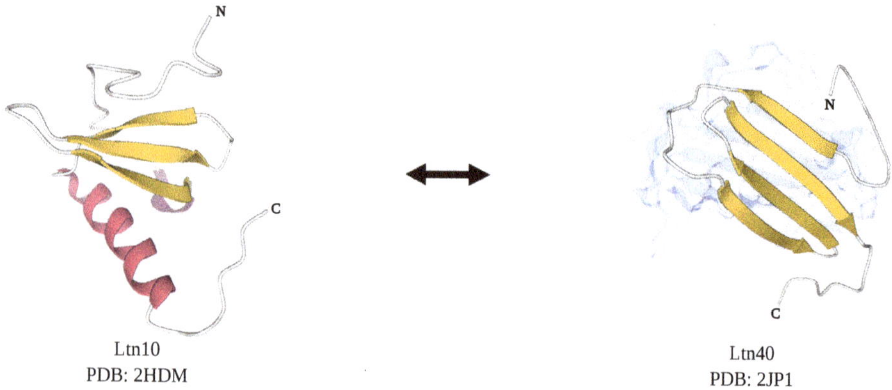

Figure 3. Structures of lymphotactin, a metamorphic protein that is also a moonlighting protein. There are two tertiary folds for lymphotactin, Ltn10 and Ltn40. Ltn10 has a classical chemokine fold with a mix of alpha-helix and beta-sheet (PDB ID: 2HDM [134]). Ltn40 possesses a dimeric form with each subunit composed of mainly beta-sheets (PDB ID: 2JP1 [125]). Note: This figure was "Created with BioRender.com", and the visualizations of the protein structures were created with Mol* [36] on the RCSB PDB website (rcsb.org) [37,38].

In addition, CLIC1 is a metamorphic protein with two native folded states and also a moonlighting protein with two different functions [136]. In an oligomeric form, CLIC1 functions as a transmembrane chloride ion channel with its *N*-terminus folded in an all-alpha native state. In a monomeric and soluble form, CLIC1 functions as an oxidoreductase with a transformed *N*-terminus that is a mixed structure containing an alpha helix and a beta sheet [137,138].

2.4.3. Morpheein Proteins

The subunits of morpheeins form a multimer that can disassemble, change conformation (without refolding), and reassemble into a different multimer [139].

Porphobilinogen synthase (also known as delta-aminolevulinic acid dehydratase) is the prototype of a morpheein. It has two oligomeric states that correlate with different levels of enzyme activity and binding of allosteric effectors [140]. An octamer can disassemble into dimers. While part of a dimer, domains within the subunits can shift in their relative positions to result in subunits with a different conformation that can then assemble into a hexamer (Figure 4). While these different homomultimers vary in their level of activity of one function, porphobilinogen synthase enzyme activity, the protein is also a moonlighting protein because it has a second function in which it binds to and inhibits the proteasome [141,142].

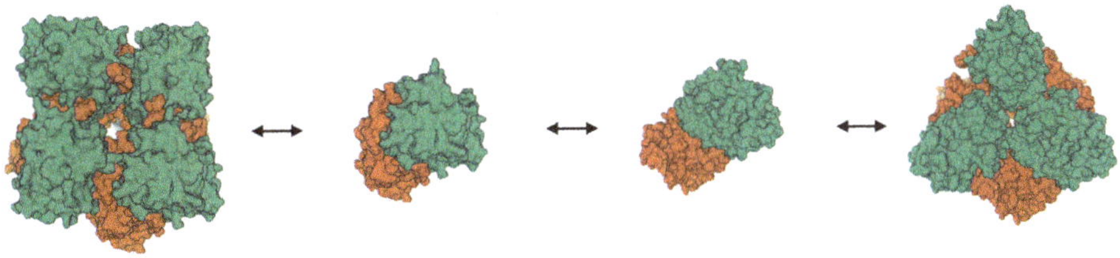

Octamer
High activity
PDB: 1I8J

Dimer
PDB: 1I8J

Dimer
PDB: 1PV8

Hexamer
Low activity
PDB: 1PV8

Figure 4. Structures of porphobilinogen synthase, which is both a morpheein and a moonlighting protein. Porphobilinogen synthase can form two homo-multimers, a low activity hexamer (PDB ID: 1PV8 [143]) and a high activity octamer (PDB ID: 1I8J [144]). The two multimers can interconvert through two homo-dimers, with different subunit conformations. In addition to its catalytic function, porphobilinogen synthase has a second function as an inhibitor of the proteasome. Note: This figure was "Created with BioRender.com", and the visualizations of the protein structures were created with Mol* [36] on the RCSB PDB website (rcsb.org) [37,38].

The ebolavirus VP40 is a morpheein and moonlighting protein that has three different functions in the virus life cycle, and each of these functions corresponds to a different arrangement of subunits [145]. An octameric ring structure binds to viral RNA to regulate its transcription while in host cells. A butterfly-shaped dimer moves to the host cell's plasma membrane. Then a linear hexameric form assembles into a larger structure that is needed for budding.

3. Moonlighting Proteins in Cellular Complexity

The variety of functions and combinations of functions of moonlighting proteins contribute to the complexity of cellular metabolism. Protein function, and in many cases the structure, is dependent on cellular factors that can vary due to intracellular conditions and the extracellular environment, and the output can change in a dynamic way. The examples given above are just some of the known factors affecting the functions of moonlighting proteins, and one protein often responds to multiple signals or combinations of signals. Some switches in function are reversible, others are not. Within a single cell, some copies of a moonlighting protein can be performing one function, some another, and some both simultaneously, depending on the protein, the cell type, and on the individual cell's metabolic state and environmental conditions.

These factors that make our understanding of the cell difficult are valuable to the cell because they help enable dynamic responses to fluctuations in conditions within the cell and in its environment. Because the function of moonlighting proteins can depend on multiple factors, they can also be components of controllable cellular responses and be involved in processing information. Some moonlighting proteins also help regulate the level of activity of other proteins in the cell, for example by an enzyme with a second function in a cell signaling pathway or as translation or transcription factors. Because moonlighting proteins can have different activities in different cell types, they also contribute to different cell types having different phenotypes with specialized functions.

In addition to actually contributing to the complexity of the many interconnected pathways and processes in the cell, the ability of a protein to perform two different functions adds to the "fuzziness" of our inability to fully understand, predict, and model the activities in a cell and how they interact and are regulated. There are many things we still don't know about moonlighting proteins. First, we don't know how many proteins are moonlighting proteins. Many protein functions were found by serendipity and researchers are often looking for one type of function when they study a protein, not all of the functions that might be there. There are also many proteins identified through sequencing projects for which we don't know any functions. We also don't completely understand the triggers and mechanisms for switching between different functions—the cellular conditions, ligands, protein–protein interactions, conformational changes, PTMs, etc. involved and how all the different triggers combine.

Our understanding of moonlighting proteins and our ability to predict which proteins are moonlighting proteins and what are their functions is also complicated because a protein can have one second function, but a homologous protein can have a different second function, for example the cytoplasmic and mitochondrial aconitases mentioned above. Leucyl-tRNA synthetase is another example. It is an enzyme that attaches leucine to tRNA, but it has additional functions that vary in yeast and humans. In the yeast *Saccharomyces cerevisiae*, leucyl-tRNA synthetase is involved in intron splicing in RNA [146]. However, in humans it is involved in cell signaling, where it senses the cellular leucine concentration and binds to and activates Rag GTPase, leading to the activation of mTORC1 (mammalian target of rapamycin complex 1) [147].

In fact, a family of homologous enzymes can include enzymes with the canonical catalytic function of the protein family, moonlighting proteins with different combinations of catalytic and non-catalytic functions, as well as enzymes with variations on the canonical catalytic function (i.e., different substrates or chemical reaction catalyzed) and even pseudoenzymes, which can resemble active enzymes but have no catalytic function [148,149]. Though noncatalytic, pseudoenzymes play important roles in regulating the activity of their catalytic homologues, facilitating the assembly of scaffolding complexes and coordinating transcription and translation [150]. The loss of the catalytic functions can be attributed to a variety of aspects, such as the loss of essential amino acid residues needed for catalysis in the active site, a blockage of the entrance to the active site, and mutations of the amino acids involved in binding the substrate [151,152]. The first reported pseudoenzyme was alpha-lactalbumin, which is homologous to the enzyme lysozyme but does not have catalytic activity [153,154]. Instead, alpha-lactalbumin is a component of lactose synthase, serving as a regulatory subunit that increases the substrate binding affinity for the catalytic subunit of the enzyme [155]. Another example is found in the argininosuccinate lyase protein family. The canonical argininosuccinate lyase enzymes catalyze the breakdown of argininosuccinate into arginine and fumarate. Delta1-crystallin is a pseudoenzyme member of the family with a function as a structural protein in the lens of the eye in birds and reptiles. Another member of the protein family, delta2-crystallin, shares 94% amino acid sequence identity with the delta1-crystallin and is both a crystallin and a catalytically active argininosuccinate lyase that catalyzes the breakdown of argininosuccinate into arginine and fumarate [156,157].

4. Conclusions

The variability in the functions of moonlighting proteins, including some intrinsically disordered, metamorphic and morpheein proteins, contribute immensely to the fuzziness concept of cellular metabolism as described by Gentili [6]. Many types of proteins have multiple functions that interact in a complex pattern of interacting pathways and processes. As cellular conditions change due to metabolism and environmental conditions, the functions of these proteins change, resulting in different combinations of interactions and processes. The fuzziness concept also represents our limited understanding of these players and our inability to fully predict and model their actions and interactions. We don't yet know how many proteins are moonlighting proteins or the full complement of their functions, and in many cases, we also don't know the cellular factors that affect their functions. Moreover, while many advances have been made in our ability to predict protein functions, the variety of functions found even among homologous proteins adds to the fuzziness of our predictions.

Author Contributions: Writing—original draft preparation, H.L. and C.J.J.; writing—review and editing, H.L. and C.J.J. All authors have read and agreed to the published version of the manuscript.

Funding: Research on moonlighting proteins in the Jeffery Lab is supported by an award from the University of Illinois Cancer Center.

Conflicts of Interest: The authors declare no conflict of interest.

References

1. Jeffery, C.J. Moonlighting proteins. *Trends Biochem. Sci.* **1999**, *24*, 8–11. [CrossRef]
2. Jeffery, C.J. Moonlighting proteins: Old proteins learning new tricks. *Trends Genet.* **2003**, *19*, 415–417. [CrossRef]
3. Jeffery, C.J. Protein moonlighting: What is it, and why is it important? *Philos. Trans. R. Soc. Lond. B Biol. Sci.* **2018**, *373*. [CrossRef] [PubMed]
4. Mani, M.; Chen, C.; Amblee, V.; Liu, H.; Mathur, T.; Zwicke, G.; Zabad, S.; Patel, B.; Thakkar, J.; Jeffery, C.J. MoonProt: A database for proteins that are known to moonlight. *Nucleic Acids Res.* **2015**, *43*, D277–D282. [CrossRef] [PubMed]
5. Chen, C.; Zabad, S.; Liu, H.; Wang, W.; Jeffery, C. MoonProt 2.0: An expansion and update of the moonlighting proteins database. *Nucleic Acids Res.* **2018**, *46*, D640–D644. [CrossRef]
6. Gentili, P.L. The Fuzziness of the Molecular World and Its Perspectives. *Molecules* **2018**, *23*, 2074. [CrossRef] [PubMed]
7. Fuxreiter, M. Towards a Stochastic Paradigm: From Fuzzy Ensembles to Cellular Functions. *Molecules* **2018**, *23*, 3008. [CrossRef] [PubMed]
8. Henderson, B.; Martin, A. Bacterial virulence in the moonlight: Multitasking bacterial moonlighting proteins are virulence determinants in infectious disease. *Infect. Immun.* **2011**, *79*, 3476–3491. [CrossRef] [PubMed]
9. Jeffery, C. Intracellular proteins moonlighting as bacterial adhesion factors. *Aims Microbiol.* **2018**, *4*, 362–376. [CrossRef]
10. Wang, W.; Jeffery, C.J. An analysis of surface proteomics results reveals novel candidates for intracellular/surface moonlighting proteins in bacteria. *Mol. Biosyst.* **2016**, *12*, 1420–1431. [CrossRef]
11. Kainulainen, V.; Korhonen, T.K. Dancing to another tune-adhesive moonlighting proteins in bacteria. *Biology* **2014**, *3*, 178–204. [CrossRef] [PubMed]
12. Sha, J.; Erova, T.E.; Alyea, R.A.; Wang, S.; Olano, J.P.; Pancholi, V.; Chopra, A.K. Surface-expressed enolase contributes to the pathogenesis of clinical isolate SSU of *Aeromonas hydrophila*. *J. Bacteriol.* **2009**, *191*, 3095–3107. [CrossRef] [PubMed]
13. Agarwal, S.; Kulshreshtha, P.; Bambah Mukku, D.; Bhatnagar, R. α-Enolase binds to human plasminogen on the surface of *Bacillus anthracis*. *Biochim. Biophys. Acta* **2008**, *1784*, 986–994. [CrossRef] [PubMed]
14. Knaust, A.; Weber, M.V.; Hammerschmidt, S.; Bergmann, S.; Frosch, M.; Kurzai, O. Cytosolic proteins contribute to surface plasminogen recruitment of *Neisseria meningitidis*. *J. Bacteriol.* **2007**, *189*, 3246–3255. [CrossRef]
15. Kolberg, J.; Aase, A.; Bergmann, S.; Herstad, T.K.; Rodal, G.; Frank, R.; Rohde, M.; Hammerschmidt, S. *Streptococcus pneumoniae* enolase is important for plasminogen binding despite low abundance of enolase protein on the bacterial cell surface. *Microbiology* **2006**, *152*, 1307–1317. [CrossRef]
16. Mundodi, V.; Kucknoor, A.S.; Alderete, J.F. Immunogenic and plasminogen-binding surface-associated alpha-enolase of Trichomonas vaginalis. *Infect. Immun.* **2008**, *76*, 523–531. [CrossRef]
17. Antikainen, J.; Kuparinen, V.; Lahteenmaki, K.; Korhonen, T.K. Enolases from Gram-positive bacterial pathogens and commensal lactobacilli share functional similarity in virulence-associated traits. *Fems Immunol. Med. Microbiol.* **2007**, *51*, 526–534. [CrossRef]
18. Collen, D.; Verstraete, M. Molecular biology of human plasminogen. II. Metabolism in physiological and some pathological conditions in man. *Thromb. Diath. Haemorrh.* **1975**, *34*, 403–408.
19. Dano, K.; Andreasen, P.A.; Grondahl-Hansen, J.; Kristensen, P.; Nielsen, L.S.; Skriver, L. Plasminogen activators, tissue degradation, and cancer. *Adv. Cancer Res.* **1985**, *44*, 139–266. [CrossRef]
20. Esgleas, M.; Li, Y.; Hancock, M.A.; Harel, J.; Dubreuil, J.D.; Gottschalk, M. Isolation and characterization of alpha-enolase, a novel fibronectin-binding protein from *Streptococcus suis*. *Microbiology* **2008**, *154*, 2668–2679. [CrossRef]
21. Carneiro, C.R.; Postol, E.; Nomizo, R.; Reis, L.F.; Brentani, R.R. Identification of enolase as a laminin-binding protein on the surface of Staphylococcus aureus. *Microbes Infect.* **2004**, *6*, 604–608. [CrossRef] [PubMed]
22. Kinoshita, H.; Uchida, H.; Kawai, Y.; Kawasaki, T.; Wakahara, N.; Matsuo, H.; Watanabe, M.; Kitazawa, H.; Ohnuma, S.; Miura, K.; et al. Cell surface Lactobacillus plantarum LA 318 glyceraldehyde-3-phosphate dehydrogenase (GAPDH) adheres to human colonic mucin. *J. Appl. Microbiol.* **2008**, *104*, 1667–1674. [CrossRef]

23. Patel, D.K.; Shah, K.R.; Pappachan, A.; Gupta, S.; Singh, D.D. Cloning, expression and characterization of a mucin-binding GAPDH from Lactobacillus acidophilus. *Int. J. Biol. Macromol.* **2016**, *91*, 338–346. [CrossRef] [PubMed]
24. Lottenberg, R.; Broder, C.C.; Boyle, M.D.; Kain, S.J.; Schroeder, B.L.; Curtiss, R., 3rd. Cloning, sequence analysis, and expression in Escherichia coli of a streptococcal plasmin receptor. *J. Bacteriol.* **1992**, *174*, 5204–5210. [CrossRef] [PubMed]
25. Pancholi, V.; Fischetti, V.A. A major surface protein on group a streptococci is a glyceraldehyde-3-phosphate-dehydrogenase with multiple binding activity. *J. Exp. Med.* **1992**, *176*, 415–426. [CrossRef]
26. Pancholi, V.; Fischetti, V.A. Glyceraldehyde-3-phosphate dehydrogenase on the surface of group A streptococci is also an ADP-ribosylating enzyme. *Proc. Natl. Acad. Sci. USA* **1993**, *90*, 8154–8158. [CrossRef] [PubMed]
27. Terao, Y.; Yamaguchi, M.; Hamada, S.; Kawabata, S. Multifunctional glyceraldehyde-3-phosphate dehydrogenase of *Streptococcus pyogenes* is essential for evasion from neutrophils. *J. Biol. Chem.* **2006**, *281*, 14215–14223. [CrossRef]
28. Madureira, P.; Baptista, M.; Vieira, M.; Magalhaes, V.; Camelo, A.; Oliveira, L.; Ribeiro, A.; Tavares, D.; Trieu-Cuot, P.; Vilanova, M.; et al. *Streptococcus agalactiae* GAPDH is a virulence-associated immunomodulatory protein. *J. Immunol.* **2007**, *178*, 1379–1387. [CrossRef]
29. Xolalpa, W.; Vallecillo, A.J.; Lara, M.; Mendoza-Hernandez, G.; Comini, M.; Spallek, R.; Singh, M.; Espitia, C. Identification of novel bacterial plasminogen-binding proteins in the human pathogen Mycobacterium tuberculosis. *Proteomics* **2007**, *7*, 3332–3341. [CrossRef]
30. Wang, Y.; Kelly, C.G.; Karttunen, J.T.; Whittall, T.; Lehner, P.J.; Duncan, L.; MacAry, P.; Younson, J.S.; Singh, M.; Oehlmann, W.; et al. CD40 is a cellular receptor mediating mycobacterial heat shock protein 70 stimulation of CC-chemokines. *Immunity* **2001**, *15*, 971–983. [CrossRef]
31. Ehinger, S.; Schubert, W.D.; Bergmann, S.; Hammerschmidt, S.; Heinz, D.W. Plasmin(ogen)-binding alpha-enolase from *Streptococcus pneumoniae*: Crystal structure and evaluation of plasmin(ogen)-binding sites. *J. Mol. Biol.* **2004**, *343*, 997–1005. [CrossRef] [PubMed]
32. Dupuy, J.; Volbeda, A.; Carpentier, P.; Darnault, C.; Moulis, J.M.; Fontecilla-Camps, J.C. Crystal structure of human iron regulatory protein 1 as cytosolic aconitase. *Structure* **2006**, *14*, 129–139. [CrossRef]
33. Walden, W.E.; Selezneva, A.I.; Dupuy, J.; Volbeda, A.; Fontecilla-Camps, J.C.; Theil, E.C.; Volz, K. Structure of dual function iron regulatory protein 1 complexed with ferritin IRE-RNA. *Science* **2006**, *314*, 1903–1908. [CrossRef] [PubMed]
34. Kim, K.H.; Lee, W.; Kim, E.E. Crystal structures of human peroxiredoxin 6 in different oxidation states. *Biochem. Biophys. Res. Commun.* **2016**, *477*, 717–722. [CrossRef] [PubMed]
35. Teixeira, F.; Tse, E.; Castro, H.; Makepeace, K.A.T.; Meinen, B.A.; Borchers, C.H.; Poole, L.B.; Bardwell, J.C.; Tomas, A.M.; Southworth, D.R.; et al. Chaperone activation and client binding of a 2-cysteine peroxiredoxin. *Nat. Commun.* **2019**, *10*, 659. [CrossRef] [PubMed]
36. Sehnal, D.; Rose, A.S.; Kovca, J.; Burley, S.K.; Velankar, S. Mol*: Towards a common library and tools for web molecular graphics. In *Workshop on Molecular Graphics and Visual Analysis of Molecular Data*; Byška, J., Krone, M., Sommer, B., Eds.; The Eurographics Association: Aire-la-Ville, Switzerland, 2018; pp. 29–33. [CrossRef]
37. Berman, H.M.; Westbrook, J.; Feng, Z.; Gilliland, G.; Bhat, T.N.; Weissig, H.; Shindyalov, I.N.; Bourne, P.E. The Protein Data Bank. *Nucleic Acids Res.* **2000**, *28*, 235–242. [CrossRef]
38. Burley, S.K.; Berman, H.M.; Bhikadiya, C.; Bi, C.; Chen, L.; Di Costanzo, L.; Christie, C.; Dalenberg, K.; Duarte, J.M.; Dutta, S.; et al. RCSB Protein Data Bank: Biological macromolecular structures enabling research and education in fundamental biology, biomedicine, biotechnology and energy. *Nucleic Acids Res.* **2019**, *47*, D464–D474. [CrossRef]
39. Artymiuk, P.J.; Green, J. The double life of aconitase. *Structure* **2006**, *14*, 2–4. [CrossRef]
40. Banerjee, S.; Nandyala, A.K.; Raviprasad, P.; Ahmed, N.; Hasnain, S.E. Iron-dependent RNA-binding activity of *Mycobacterium tuberculosis* aconitase. *J. Bacteriol.* **2007**, *189*, 4046–4052. [CrossRef]
41. Kennedy, M.C.; Mende-Mueller, L.; Blondin, G.A.; Beinert, H. Purification and characterization of cytosolic aconitase from beef liver and its relationship to the iron-responsive element binding protein. *Proc. Natl. Acad. Sci. USA* **1992**, *89*, 11730–11734. [CrossRef]
42. Howard, P.K.; Shaw, J.; Otsuka, A.J. Nucleotide sequence of the birA gene encoding the biotin operon repressor and biotin holoenzyme synthetase functions of Escherichia coli. *Gene* **1985**, *35*, 321–331. [CrossRef]

43. Barker, D.F.; Campbell, A.M. Genetic and biochemical characterization of the birA gene and its product: Evidence for a direct role of biotin holoenzyme synthetase in repression of the biotin operon in Escherichia coli. *J. Mol. Biol.* **1981**, *146*, 469–492. [CrossRef]
44. Wan, F.; Anderson, D.E.; Barnitz, R.A.; Snow, A.; Bidere, N.; Zheng, L.; Hegde, V.; Lam, L.T.; Staudt, L.M.; Levens, D.; et al. Ribosomal protein S3: A KH domain subunit in NF-kappaB complexes that mediates selective gene regulation. *Cell* **2007**, *131*, 927–939. [CrossRef]
45. Malygin, A.A.; Parakhnevitch, N.M.; Ivanov, A.V.; Eperon, I.C.; Karpova, G.G. Human ribosomal protein S13 regulates expression of its own gene at the splicing step by a feedback mechanism. *Nucleic Acids Res.* **2007**, *35*, 6414–6423. [CrossRef]
46. Fewell, S.W.; Woolford, J.L., Jr. Ribosomal protein S14 of *Saccharomyces cerevisiae* regulates its expression by binding to RPS14B pre-mRNA and to 18S rRNA. *Mol. Cell Biol.* **1999**, *19*, 826–834. [CrossRef] [PubMed]
47. Presutti, C.; Ciafre, S.A.; Bozzoni, I. The ribosomal protein L2 in S. cerevisiae controls the level of accumulation of its own mRNA. *EMBO J.* **1991**, *10*, 2215–2221. [CrossRef]
48. Zengel, J.M.; Lindahl, L. A hairpin structure upstream of the terminator hairpin required for ribosomal protein L4-mediated attenuation control of the S10 operon of Escherichia coli. *J. Bacteriol.* **1996**, *178*, 2383–2387. [CrossRef]
49. Horn, H.F.; Vousden, K.H. Cooperation between the ribosomal proteins L5 and L11 in the p53 pathway. *Oncogene* **2008**, *27*, 5774–5784. [CrossRef]
50. Kuhn, J.F.; Tran, E.J.; Maxwell, E.S. Archaeal ribosomal protein L7 is a functional homolog of the eukaryotic 15.5kD/Snu13p snoRNP core protein. *Nucleic Acids Res.* **2002**, *30*, 931–941. [CrossRef]
51. Lohrum, M.A.; Ludwig, R.L.; Kubbutat, M.H.; Hanlon, M.; Vousden, K.H. Regulation of HDM2 activity by the ribosomal protein L11. *Cancer Cell* **2003**, *3*, 577–587. [CrossRef]
52. Mazumder, B.; Sampath, P.; Seshadri, V.; Maitra, R.K.; DiCorleto, P.E.; Fox, P.L. Regulated release of L13a from the 60S ribosomal subunit as a mechanism of transcript-specific translational control. *Cell* **2003**, *115*, 187–198. [CrossRef]
53. Dai, M.S.; Zeng, S.X.; Jin, Y.; Sun, X.X.; David, L.; Lu, H. Ribosomal protein L23 activates p53 by inhibiting MDM2 function in response to ribosomal perturbation but not to translation inhibition. *Mol. Cell Biol.* **2004**, *24*, 7654–7668. [CrossRef] [PubMed]
54. Takagi, M.; Absalon, M.J.; McLure, K.G.; Kastan, M.B. Regulation of p53 translation and induction after DNA damage by ribosomal protein L26 and nucleolin. *Cell* **2005**, *123*, 49–63. [CrossRef] [PubMed]
55. Wilson, D.M., 3rd; Deutsch, W.A.; Kelley, M.R. Drosophila ribosomal protein S3 contains an activity that cleaves DNA at apurinic/apyrimidinic sites. *J. Biol. Chem.* **1994**, *269*, 25359–25364. [PubMed]
56. Chen, X.J.; Wang, X.; Kaufman, B.A.; Butow, R.A. Aconitase couples metabolic regulation to mitochondrial DNA maintenance. *Science* **2005**, *307*, 714–717. [CrossRef]
57. Jang, H.H.; Lee, K.O.; Chi, Y.H.; Jung, B.G.; Park, S.K.; Park, J.H.; Lee, J.R.; Lee, S.S.; Moon, J.C.; Yun, J.W.; et al. Two enzymes in one; two yeast peroxiredoxins display oxidative stress-dependent switching from a peroxidase to a molecular chaperone function. *Cell* **2004**, *117*, 625–635. [CrossRef]
58. Krojer, T.; Garrido-Franco, M.; Huber, R.; Ehrmann, M.; Clausen, T. Crystal structure of DegP (HtrA) reveals a new protease-chaperone machine. *Nature* **2002**, *416*, 455–459. [CrossRef]
59. Jeffery, C.J. An enzyme in the test tube, and a transcription factor in the cell: Moonlighting proteins and cellular factors that affect their behavior. *Protein Sci.* **2019**, *28*, 1233–1238. [CrossRef]
60. Romero, P.; Obradovic, Z.; Dunker, A.K. Natively disordered proteins: Functions and predictions. *Appl. Bioinform.* **2004**, *3*, 105–113. [CrossRef]
61. Dyson, H.J. Making Sense of Intrinsically Disordered Proteins. *Biophys. J.* **2016**, *110*, 1013–1016. [CrossRef]
62. Dunker, A.K.; Lawson, J.D.; Brown, C.J.; Williams, R.M.; Romero, P.; Oh, J.S.; Oldfield, C.J.; Campen, A.M.; Ratliff, C.M.; Hipps, K.W.; et al. Intrinsically disordered protein. *J. Mol. Graph. Model.* **2001**, *19*, 26–59. [CrossRef]
63. Tompa, P.; Szasz, C.; Buday, L. Structural disorder throws new light on moonlighting. *Trends Biochem. Sci.* **2005**, *30*, 484–489. [CrossRef] [PubMed]
64. Wright, P.E.; Dyson, H.J. Intrinsically unstructured proteins: Re-assessing the protein structure-function paradigm. *J. Mol. Biol.* **1999**, *293*, 321–331. [CrossRef] [PubMed]
65. Dunker, A.K.; Brown, C.J.; Lawson, J.D.; Iakoucheva, L.M.; Obradovic, Z. Intrinsic disorder and protein function. *Biochemistry* **2002**, *41*, 6573–6582. [CrossRef]

66. Hatos, A.; Hajdu-Soltesz, B.; Monzon, A.M.; Palopoli, N.; Alvarez, L.; Aykac-Fas, B.; Bassot, C.; Benitez, G.I.; Bevilacqua, M.; Chasapi, A.; et al. DisProt: Intrinsic protein disorder annotation in 2020. *Nucleic Acids Res.* **2020**, *48*, D269–D276. [CrossRef]
67. Vogelstein, B.; Lane, D.; Levine, A.J. Surfing the p53 network. *Nature* **2000**, *408*, 307–310. [CrossRef]
68. Green, D.R.; Kroemer, G. Cytoplasmic functions of the tumour suppressor p53. *Nature* **2009**, *458*, 1127–1130. [CrossRef]
69. Zhou, J.; Ahn, J.; Wilson, S.H.; Prives, C. A role for p53 in base excision repair. *EMBO J.* **2001**, *20*, 914–923. [CrossRef]
70. Clore, G.M.; Omichinski, J.G.; Sakaguchi, K.; Zambrano, N.; Sakamoto, H.; Appella, E.; Gronenborn, A.M. High-resolution structure of the oligomerization domain of p53 by multidimensional NMR. *Science* **1994**, *265*, 386–391. [CrossRef]
71. Wells, M.; Tidow, H.; Rutherford, T.J.; Markwick, P.; Jensen, M.R.; Mylonas, E.; Svergun, D.I.; Blackledge, M.; Fersht, A.R. Structure of tumor suppressor p53 and its intrinsically disordered *N*-terminal transactivation domain. *Proc. Natl. Acad. Sci. USA* **2008**, *105*, 5762–5767. [CrossRef]
72. Bell, S.; Klein, C.; Muller, L.; Hansen, S.; Buchner, J. p53 contains large unstructured regions in its native state. *J. Mol. Biol.* **2002**, *322*, 917–927. [CrossRef]
73. Safer, D.; Elzinga, M.; Nachmias, V.T. Thymosin beta 4 and Fx, an actin-sequestering peptide, are indistinguishable. *J. Biol. Chem.* **1991**, *266*, 4029–4032. [PubMed]
74. Cassimeris, L.; Safer, D.; Nachmias, V.T.; Zigmond, S.H. Thymosin beta 4 sequesters the majority of G-actin in resting human polymorphonuclear leukocytes. *J. Cell Biol.* **1992**, *119*, 1261–1270. [CrossRef] [PubMed]
75. Grant, D.S.; Rose, W.; Yaen, C.; Goldstein, A.; Martinez, J.; Kleinman, H. Thymosin beta4 enhances endothelial cell differentiation and angiogenesis. *Angiogenesis* **1999**, *3*, 125–135. [CrossRef] [PubMed]
76. Smart, N.; Rossdeutsch, A.; Riley, P.R. Thymosin beta4 and angiogenesis: Modes of action and therapeutic potential. *Angiogenesis* **2007**, *10*, 229–241. [CrossRef] [PubMed]
77. Philp, D.; Kleinman, H.K. Animal studies with thymosin beta, a multifunctional tissue repair and regeneration peptide. *Ann. NY Acad. Sci.* **2010**, *1194*, 81–86. [CrossRef] [PubMed]
78. Young, J.D.; Lawrence, A.J.; MacLean, A.G.; Leung, B.P.; McInnes, I.B.; Canas, B.; Pappin, D.J.; Stevenson, R.D. Thymosin beta 4 sulfoxide is an anti-inflammatory agent generated by monocytes in the presence of glucocorticoids. *Nat. Med.* **1999**, *5*, 1424–1427. [CrossRef] [PubMed]
79. Safer, D.; Sosnick, T.R.; Elzinga, M. Thymosin beta 4 binds actin in an extended conformation and contacts both the barbed and pointed ends. *Biochemistry* **1997**, *36*, 5806–5816. [CrossRef]
80. Domanski, M.; Hertzog, M.; Coutant, J.; Gutsche-Perelroizen, I.; Bontems, F.; Carlier, M.F.; Guittet, E.; van Heijenoort, C. Coupling of folding and binding of thymosin beta4 upon interaction with monomeric actin monitored by nuclear magnetic resonance. *J. Biol. Chem.* **2004**, *279*, 23637–23645. [CrossRef]
81. Bock-Marquette, I.; Saxena, A.; White, M.D.; Dimaio, J.M.; Srivastava, D. Thymosin beta4 activates integrin-linked kinase and promotes cardiac cell migration, survival and cardiac repair. *Nature* **2004**, *432*, 466–472. [CrossRef]
82. Ho, J.H.; Tseng, K.C.; Ma, W.H.; Chen, K.H.; Lee, O.K.; Su, Y. Thymosin beta-4 upregulates anti-oxidative enzymes and protects human cornea epithelial cells against oxidative damage. *Br. J. Ophthalmol.* **2008**, *92*, 992–997. [CrossRef] [PubMed]
83. Tantos, A.; Szabo, B.; Lang, A.; Varga, Z.; Tsylonok, M.; Bokor, M.; Verebelyi, T.; Kamasa, P.; Tompa, K.; Perczel, A.; et al. Multiple fuzzy interactions in the moonlighting function of thymosin-beta4. *Intrinsically Disord Proteins* **2013**, *1*, e26204. [CrossRef] [PubMed]
84. Czisch, M.; Schleicher, M.; Horger, S.; Voelter, W.; Holak, T.A. Conformation of thymosin beta 4 in water determined by NMR spectroscopy. *Eur. J. Biochem.* **1993**, *218*, 335–344. [CrossRef] [PubMed]
85. Hope, I.A.; Struhl, K. Functional dissection of a eukaryotic transcriptional activator protein, GCN4 of yeast. *Cell* **1986**, *46*, 885–894. [CrossRef]
86. Nikolaev, Y.; Deillon, C.; Hoffmann, S.R.; Bigler, L.; Friess, S.; Zenobi, R.; Pervushin, K.; Hunziker, P.; Gutte, B. The leucine zipper domains of the transcription factors GCN4 and c-Jun have ribonuclease activity. *PLoS ONE* **2010**, *5*, e10765. [CrossRef]
87. Weiss, M.A.; Ellenberger, T.; Wobbe, C.R.; Lee, J.P.; Harrison, S.C.; Struhl, K. Folding transition in the DNA-binding domain of GCN4 on specific binding to DNA. *Nature* **1990**, *347*, 575–578. [CrossRef]

88. Leon-Del-Rio, A.; Valadez-Graham, V.; Gravel, R.A. Holocarboxylase Synthetase: A Moonlighting Transcriptional Coregulator of Gene Expression and a Cytosolic Regulator of Biotin Utilization. *Annu. Rev. Nutr.* **2017**, *37*, 207–223. [CrossRef]
89. Wilson, K.P.; Shewchuk, L.M.; Brennan, R.G.; Otsuka, A.J.; Matthews, B.W. Escherichia coli biotin holoenzyme synthetase/bio repressor crystal structure delineates the biotin- and DNA-binding domains. *Proc. Natl. Acad. Sci. USA* **1992**, *89*, 9257–9261. [CrossRef]
90. Merenmies, J.; Pihlaskari, R.; Laitinen, J.; Wartiovaara, J.; Rauvala, H. 30-kDa heparin-binding protein of brain (amphoterin) involved in neurite outgrowth. Amino acid sequence and localization in the filopodia of the advancing plasma membrane. *J. Biol. Chem.* **1991**, *266*, 16722–16729.
91. Hardman, C.H.; Broadhurst, R.W.; Raine, A.R.; Grasser, K.D.; Thomas, J.O.; Laue, E.D. Structure of the A-domain of HMG1 and its interaction with DNA as studied by heteronuclear three- and four-dimensional NMR spectroscopy. *Biochemistry* **1995**, *34*, 16596–16607. [CrossRef]
92. Stott, K.; Watson, M.; Howe, F.S.; Grossmann, J.G.; Thomas, J.O. Tail-mediated collapse of HMGB1 is dynamic and occurs via differential binding of the acidic tail to the A and B domains. *J. Mol. Biol.* **2010**, *403*, 706–722. [CrossRef] [PubMed]
93. Brunagel, G.; Shah, U.; Schoen, R.E.; Getzenberg, R.H. Identification of calreticulin as a nuclear matrix protein associated with human colon cancer. *J. Cell Biochem.* **2003**, *89*, 238–243. [CrossRef] [PubMed]
94. Goicoechea, S.; Orr, A.W.; Pallero, M.A.; Eggleton, P.; Murphy-Ullrich, J.E. Thrombospondin mediates focal adhesion disassembly through interactions with cell surface calreticulin. *J. Biol. Chem.* **2000**, *275*, 36358–36368. [CrossRef] [PubMed]
95. Bouvier, M.; Stafford, W.F. Probing the three-dimensional structure of human calreticulin. *Biochemistry* **2000**, *39*, 14950–14959. [CrossRef] [PubMed]
96. Yates, J.L.; Nomura, M.E. coli ribosomal protein L4 is a feedback regulatory protein. *Cell* **1980**, *21*, 517–522. [CrossRef]
97. Timsit, Y.; Acosta, Z.; Allemand, F.; Chiaruttini, C.; Springer, M. The role of disordered ribosomal protein extensions in the early steps of eubacterial 50 S ribosomal subunit assembly. *Int. J. Mol. Sci.* **2009**, *10*, 817–834. [CrossRef]
98. Wohlschlegel, J.A.; Dwyer, B.T.; Dhar, S.K.; Cvetic, C.; Walter, J.C.; Dutta, A. Inhibition of eukaryotic DNA replication by geminin binding to Cdt1. *Science* **2000**, *290*, 2309–2312. [CrossRef]
99. Varma, D.; Chandrasekaran, S.; Sundin, L.J.; Reidy, K.T.; Wan, X.; Chasse, D.A.; Nevis, K.R.; DeLuca, J.G.; Salmon, E.D.; Cook, J.G. Recruitment of the human Cdt1 replication licensing protein by the loop domain of Hec1 is required for stable kinetochore-microtubule attachment. *Nat. Cell Biol.* **2012**, *14*, 593–603. [CrossRef]
100. Khayrutdinov, B.I.; Bae, W.J.; Yun, Y.M.; Lee, J.H.; Tsuyama, T.; Kim, J.J.; Hwang, E.; Ryu, K.S.; Cheong, H.K.; Cheong, C.; et al. Structure of the Cdt1 *C*-terminal domain: Conservation of the winged helix fold in replication licensing factors. *Protein Sci.* **2009**, *18*, 2252–2264. [CrossRef]
101. Coschigano, P.W.; Magasanik, B. The URE2 gene product of *Saccharomyces cerevisiae* plays an important role in the cellular response to the nitrogen source and has homology to glutathione s-transferases. *Mol. Cell Biol.* **1991**, *11*, 822–832. [CrossRef]
102. Bai, M.; Zhou, J.M.; Perrett, S. The yeast prion protein Ure2 shows glutathione peroxidase activity in both native and fibrillar forms. *J. Biol. Chem.* **2004**, *279*, 50025–50030. [CrossRef] [PubMed]
103. Pierce, M.M.; Baxa, U.; Steven, A.C.; Bax, A.; Wickner, R.B. Is the prion domain of soluble Ure2p unstructured? *Biochemistry* **2005**, *44*, 321–328. [CrossRef] [PubMed]
104. Costi, M.P.; Ferrari, S.; Venturelli, A.; Calo, S.; Tondi, D.; Barlocco, D. Thymidylate synthase structure, function and implication in drug discovery. *Curr. Med. Chem.* **2005**, *12*, 2241–2258. [CrossRef] [PubMed]
105. Chu, E.; Koeller, D.M.; Casey, J.L.; Drake, J.C.; Chabner, B.A.; Elwood, P.C.; Zinn, S.; Allegra, C.J. Autoregulation of human thymidylate synthase messenger RNA translation by thymidylate synthase. *Proc. Natl. Acad. Sci. USA* **1991**, *88*, 8977–8981. [CrossRef]
106. Phan, J.; Steadman, D.J.; Koli, S.; Ding, W.C.; Minor, W.; Dunlap, R.B.; Berger, S.H.; Lebioda, L. Structure of human thymidylate synthase suggests advantages of chemotherapy with noncompetitive inhibitors. *J. Biol. Chem.* **2001**, *276*, 14170–14177. [CrossRef]
107. Margoliash, E.; Smith, E.L.; Kreil, G.; Tuppy, H. Amino-acid sequence of horse heart cytochrome c. *Nature* **1961**, *192*, 1125–1127. [CrossRef]

108. Yu, T.; Wang, X.; Purring-Koch, C.; Wei, Y.; McLendon, G.L. A mutational epitope for cytochrome C binding to the apoptosis protease activation factor-1. *J. Biol. Chem.* **2001**, *276*, 13034–13038. [CrossRef]
109. Stellwagen, E.; Rysavy, R.; Babul, G. The conformation of horse heart apocytochrome c. *J. Biol. Chem.* **1972**, *247*, 8074–8077.
110. Lux, S.E.; John, K.M.; Kopito, R.R.; Lodish, H.F. Cloning and characterization of band 3, the human erythrocyte anion-exchange protein (AE1). *Proc. Natl. Acad. Sci. USA* **1989**, *86*, 9089–9093. [CrossRef]
111. Low, P.S.; Rathinavelu, P.; Harrison, M.L. Regulation of glycolysis via reversible enzyme binding to the membrane protein, band 3. *J. Biol. Chem.* **1993**, *268*, 14627–14631.
112. Zhang, D.; Kiyatkin, A.; Bolin, J.T.; Low, P.S. Crystallographic structure and functional interpretation of the cytoplasmic domain of erythrocyte membrane band 3. *Blood* **2000**, *96*, 2925–2933. [CrossRef]
113. Zielenski, J.; Rozmahel, R.; Bozon, D.; Kerem, B.; Grzelczak, Z.; Riordan, J.R.; Rommens, J.; Tsui, L.C. Genomic DNA sequence of the cystic fibrosis transmembrane conductance regulator (CFTR) gene. *Genomics* **1991**, *10*, 214–228. [CrossRef]
114. Stutts, M.J.; Canessa, C.M.; Olsen, J.C.; Hamrick, M.; Cohn, J.A.; Rossier, B.C.; Boucher, R.C. CFTR as a cAMP-dependent regulator of sodium channels. *Science* **1995**, *269*, 847–850. [CrossRef] [PubMed]
115. Ostedgaard, L.S.; Baldursson, O.; Vermeer, D.W.; Welsh, M.J.; Robertson, A.D. A functional R domain from cystic fibrosis transmembrane conductance regulator is predominantly unstructured in solution. *Proc. Natl. Acad. Sci. USA* **2000**, *97*, 5657–5662. [CrossRef]
116. Fuxreiter, M. Fuzziness in Protein Interactions-A Historical Perspective. *J. Mol. Biol.* **2018**, *430*, 2278–2287. [CrossRef]
117. Tompa, P.; Fuxreiter, M. Fuzzy complexes: Polymorphism and structural disorder in protein-protein interactions. *Trends Biochem. Sci.* **2008**, *33*, 2–8. [CrossRef] [PubMed]
118. Sharma, R.; Raduly, Z.; Miskei, M.; Fuxreiter, M. Fuzzy complexes: Specific binding without complete folding. *FEBS Lett.* **2015**, *589*, 2533–2542. [CrossRef] [PubMed]
119. Uversky, V.N. Supramolecular Fuzziness of Intracellular Liquid Droplets: Liquid-Liquid Phase Transitions, Membrane-Less Organelles, and Intrinsic Disorder. *Molecules* **2019**, *24*, 3265. [CrossRef]
120. Miskei, M.; Antal, C.; Fuxreiter, M. FuzDB: Database of fuzzy complexes, a tool to develop stochastic structure-function relationships for protein complexes and higher-order assemblies. *Nucleic Acids Res.* **2017**, *45*, D228–D235. [CrossRef]
121. Murzin, A.G. Biochemistry. Metamorphic proteins. *Science* **2008**, *320*, 1725–1726. [CrossRef]
122. Dishman, A.F.; Volkman, B.F. Unfolding the Mysteries of Protein Metamorphosis. *Acs. Chem. Biol.* **2018**, *13*, 1438–1446. [CrossRef] [PubMed]
123. Lella, M.; Mahalakshmi, R. Metamorphic Proteins: Emergence of Dual Protein Folds from One Primary Sequence. *Biochemistry* **2017**, *56*, 2971–2984. [CrossRef] [PubMed]
124. Goodchild, S.C.; Curmi, P.M.G.; Brown, L.J. Structural gymnastics of multifunctional metamorphic proteins. *Biophys. Rev.* **2011**, *3*, 143. [CrossRef] [PubMed]
125. Tuinstra, R.L.; Peterson, F.C.; Kutlesa, S.; Elgin, E.S.; Kron, M.A.; Volkman, B.F. Interconversion between two unrelated protein folds in the lymphotactin native state. *Proc. Natl. Acad. Sci. USA* **2008**, *105*, 5057–5062. [CrossRef]
126. Burmann, B.M.; Knauer, S.H.; Sevostyanova, A.; Schweimer, K.; Mooney, R.A.; Landick, R.; Artsimovitch, I.; Rosch, P. An alpha helix to beta barrel domain switch transforms the transcription factor RfaH into a translation factor. *Cell* **2012**, *150*, 291–303. [CrossRef]
127. Goodchild, S.C.; Howell, M.W.; Littler, D.R.; Mandyam, R.A.; Sale, K.L.; Mazzanti, M.; Breit, S.N.; Curmi, P.M.; Brown, L.J. Metamorphic response of the CLIC1 chloride intracellular ion channel protein upon membrane interaction. *Biochemistry* **2010**, *49*, 5278–5289. [CrossRef]
128. Luo, X.; Yu, H. Protein metamorphosis: The two-state behavior of Mad2. *Structure* **2008**, *16*, 1616–1625. [CrossRef]
129. Tseng, R.; Goularte, N.F.; Chavan, A.; Luu, J.; Cohen, S.E.; Chang, Y.G.; Heisler, J.; Li, S.; Michael, A.K.; Tripathi, S.; et al. Structural basis of the day-night transition in a bacterial circadian clock. *Science* **2017**, *355*, 1174–1180. [CrossRef]
130. Dai, Z.; Tonelli, M.; Markley, J.L. Metamorphic protein IscU changes conformation by cis-trans isomerizations of two peptidyl-prolyl peptide bonds. *Biochemistry* **2012**, *51*, 9595–9602. [CrossRef]

131. Lopez-Pelegrin, M.; Cerda-Costa, N.; Cintas-Pedrola, A.; Herranz-Trillo, F.; Bernado, P.; Peinado, J.R.; Arolas, J.L.; Gomis-Ruth, F.X. Multiple stable conformations account for reversible concentration-dependent oligomerization and autoinhibition of a metamorphic metallopeptidase. *Angew. Chem. Int. Ed. Engl.* **2014**, *53*, 10624–10630. [CrossRef]
132. London, R.E. HIV-1 Reverse Transcriptase: A Metamorphic Protein with Three Stable States. *Structure* **2019**, *27*, 420–426. [CrossRef] [PubMed]
133. Peterson, F.C.; Elgin, E.S.; Nelson, T.J.; Zhang, F.; Hoeger, T.J.; Linhardt, R.J.; Volkman, B.F. Identification and characterization of a glycosaminoglycan recognition element of the C chemokine lymphotactin. *J. Biol. Chem.* **2004**, *279*, 12598–12604. [CrossRef] [PubMed]
134. Tuinstra, R.L.; Peterson, F.C.; Elgin, E.S.; Pelzek, A.J.; Volkman, B.F. An engineered second disulfide bond restricts lymphotactin/XCL1 to a chemokine-like conformation with XCR1 agonist activity. *Biochemistry* **2007**, *46*, 2564–2573. [CrossRef] [PubMed]
135. Knauer, S.H.; Artsimovitch, I.; Rosch, P. Transformer proteins. *Cell Cycle* **2012**, *11*, 4289–4290. [CrossRef]
136. Al Khamici, H.; Brown, L.J.; Hossain, K.R.; Hudson, A.L.; Sinclair-Burton, A.A.; Ng, J.P.; Daniel, E.L.; Hare, J.E.; Cornell, B.A.; Curmi, P.M.; et al. Members of the chloride intracellular ion channel protein family demonstrate glutaredoxin-like enzymatic activity. *PLoS ONE* **2015**, *10*, e115699. [CrossRef]
137. Littler, D.R.; Harrop, S.J.; Fairlie, W.D.; Brown, L.J.; Pankhurst, G.J.; Pankhurst, S.; DeMaere, M.Z.; Campbell, T.J.; Bauskin, A.R.; Tonini, R.; et al. The intracellular chloride ion channel protein CLIC1 undergoes a redox-controlled structural transition. *J. Biol. Chem.* **2004**, *279*, 9298–9305. [CrossRef]
138. Warton, K.; Tonini, R.; Fairlie, W.D.; Matthews, J.M.; Valenzuela, S.M.; Qiu, M.R.; Wu, W.M.; Pankhurst, S.; Bauskin, A.R.; Harrop, S.J.; et al. Recombinant CLIC1 (NCC27) assembles in lipid bilayers via a pH-dependent two-state process to form chloride ion channels with identical characteristics to those observed in Chinese hamster ovary cells expressing CLIC1. *J. Biol. Chem.* **2002**, *277*, 26003–26011. [CrossRef]
139. Jaffe, E.K. Morpheeins—A new structural paradigm for allosteric regulation. *Trends Biochem. Sci.* **2005**, *30*, 490–497. [CrossRef]
140. Jaffe, E.K.; Lawrence, S.H. Allostery and the dynamic oligomerization of porphobilinogen synthase. *Arch. Biochem. Biophys.* **2012**, *519*, 144–153. [CrossRef]
141. Guo, G.G.; Gu, M.; Etlinger, J.D. 240-kDa proteasome inhibitor (CF-2) is identical to delta-aminolevulinic acid dehydratase. *J. Biol. Chem.* **1994**, *269*, 12399–12402.
142. Li, X.C.; Gu, M.Z.; Etlinger, J.D. Isolation and characterization of a novel endogenous inhibitor of the proteasome. *Biochemistry* **1991**, *30*, 9709–9715. [CrossRef] [PubMed]
143. Breinig, S.; Kervinen, J.; Stith, L.; Wasson, A.S.; Fairman, R.; Wlodawer, A.; Zdanov, A.; Jaffe, E.K. Control of tetrapyrrole biosynthesis by alternate quaternary forms of porphobilinogen synthase. *Nat. Struct. Biol.* **2003**, *10*, 757–763. [CrossRef] [PubMed]
144. Kervinen, J.; Jaffe, E.K.; Stauffer, F.; Neier, R.; Wlodawer, A.; Zdanov, A. Mechanistic basis for suicide inactivation of porphobilinogen synthase by 4,7-dioxosebacic acid, an inhibitor that shows dramatic species selectivity. *Biochemistry* **2001**, *40*, 8227–8236. [CrossRef]
145. Bornholdt, Z.A.; Noda, T.; Abelson, D.M.; Halfmann, P.; Wood, M.R.; Kawaoka, Y.; Saphire, E.O. Structural rearrangement of ebola virus VP40 begets multiple functions in the virus life cycle. *Cell* **2013**, *154*, 763–774. [CrossRef] [PubMed]
146. Sarkar, J.; Poruri, K.; Boniecki, M.T.; McTavish, K.K.; Martinis, S.A. Yeast mitochondrial leucyl-tRNA synthetase CP1 domain has functionally diverged to accommodate RNA splicing at expense of hydrolytic editing. *J. Biol. Chem.* **2012**, *287*, 14772–14781. [CrossRef] [PubMed]
147. Han, J.M.; Jeong, S.J.; Park, M.C.; Kim, G.; Kwon, N.H.; Kim, H.K.; Ha, S.H.; Ryu, S.H.; Kim, S. Leucyl-tRNA synthetase is an intracellular leucine sensor for the mTORC1-signaling pathway. *Cell* **2012**, *149*, 410–424. [CrossRef]
148. Eyers, P.A.; Murphy, J.M. The evolving world of pseudoenzymes: Proteins, prejudice and zombies. *BMC Biol.* **2016**, *14*, 98. [CrossRef]
149. Murphy, J.M.; Farhan, H.; Eyers, P.A. Bio-Zombie: The rise of pseudoenzymes in biology. *Biochem. Soc. Trans.* **2017**, *45*, 537–544. [CrossRef]
150. Jeffery, C.J. The demise of catalysis, but new functions arise: Pseudoenzymes as the phoenixes of the protein world. *Biochem. Soc. Trans.* **2019**, *47*, 371–379. [CrossRef]

151. Todd, A.E.; Orengo, C.A.; Thornton, J.M. Sequence and structural differences between enzyme and nonenzyme homologs. *Structure* **2002**, *10*, 1435–1451. [CrossRef]
152. Pils, B.; Schultz, J. Inactive enzyme-homologues find new function in regulatory processes. *J. Mol. Biol.* **2004**, *340*, 399–404. [CrossRef]
153. Qasba, P.K.; Safaya, S.K. Similarity of the nucleotide sequences of rat alpha-lactalbumin and chicken lysozyme genes. *Nature* **1984**, *308*, 377–380. [CrossRef] [PubMed]
154. Brew, K.; Vanaman, T.C.; Hill, R.L. Comparison of the amino acid sequence of bovine alpha-lactalbumin and hens egg white lysozyme. *J. Biol. Chem.* **1967**, *242*, 3747–3749. [PubMed]
155. Powell, J.T.; Brew, K. On the interaction of alpha-lactalbumin and galactosyltransferase during lactose synthesis. *J. Biol. Chem.* **1975**, *250*, 6337–6343. [PubMed]
156. Piatigorsky, J.; O'Brien, W.E.; Norman, B.L.; Kalumuck, K.; Wistow, G.J.; Borras, T.; Nickerson, J.M.; Wawrousek, E.F. Gene sharing by delta-crystallin and argininosuccinate lyase. *Proc. Natl. Acad. Sci. USA* **1988**, *85*, 3479–3483. [CrossRef]
157. Sampaleanu, L.M.; Vallee, F.; Slingsby, C.; Howell, P.L. Structural studies of duck delta 1 and delta 2 crystallin suggest conformational changes occur during catalysis. *Biochemistry* **2001**, *40*, 2732–2742. [CrossRef]

molecules

MDPI

Perspective

Supramolecular Fuzziness of Intracellular Liquid Droplets: Liquid–Liquid Phase Transitions, Membrane-Less Organelles, and Intrinsic Disorder

Vladimir N. Uversky [1,2,*]

[1] Department of Molecular Medicine and USF Health Byrd Alzheimer's Research Institute, Morsani College of Medicine, University of South Florida, 12901 Bruce B. Downs Blvd. MDC07, Tampa, FL 33612, USA; vuversky@health.usf.edu; Tel.: +1-813-974-5816; Fax: +1-813-974-7357

[2] Institute for Biological Instrumentation of the Russian Academy of Sciences, Pushchino 142290, Moscow Region, Russia

Academic Editor: Pier Luigi Gentili
Received: 5 August 2019; Accepted: 6 September 2019; Published: 7 September 2019

Abstract: Cells are inhomogeneously crowded, possessing a wide range of intracellular liquid droplets abundantly present in the cytoplasm of eukaryotic and bacterial cells, in the mitochondrial matrix and nucleoplasm of eukaryotes, and in the chloroplast's stroma of plant cells. These proteinaceous membrane-less organelles (PMLOs) not only represent a natural method of intracellular compartmentalization, which is crucial for successful execution of various biological functions, but also serve as important means for the processing of local information and rapid response to the fluctuations in environmental conditions. Since PMLOs, being complex macromolecular assemblages, possess many characteristic features of liquids, they represent highly dynamic (or fuzzy) protein–protein and/or protein–nucleic acid complexes. The biogenesis of PMLOs is controlled by specific intrinsically disordered proteins (IDPs) and hybrid proteins with ordered domains and intrinsically disordered protein regions (IDPRs), which, due to their highly dynamic structures and ability to facilitate multivalent interactions, serve as indispensable drivers of the biological liquid–liquid phase transitions (LLPTs) giving rise to PMLOs. In this article, the importance of the disorder-based supramolecular fuzziness for LLPTs and PMLO biogenesis is discussed.

Keywords: intrinsically disordered protein; intrinsically disordered protein region; liquid–liquid phase transition; protein–protein interaction; protein–nucleic acid interaction; proteinaceous membrane-less organelle; fuzzy complex.

---

1. Introduction to Proteinaceous Membrane-Less Organelles

It is recognized now that the cellular interior represents a highly crowded space, where various biological macromolecules (such as nucleic acids, polysaccharides, proteins, and ribonucleoproteins) occupy 5–40% of the cellular volume, and where the total concentration of these biological macromolecules can be as high as 80–400 mg/mL [1,2], with the total intracellular concentration of protein being expected to be up to 300 mg/mL, while the RNA levels can range from 20–100 mg/mL [3]. Importantly, recent studies revealed that all these biomacromolecules are distributed within a cell in a highly inhomogeneous manner, often forming different intracellular bodies or intracellular liquid droplets, which are known by different names, such as cellular (or nuclear) micro-domains, cellular (or nuclear, or mitochondrial) subdomains, intracellular (or intranuclear, or intramitochondrial, or intrachloroplast) bodies, non-membranous cytoplasmic (or nucleoplasmic) granules, and proteinaceous membrane-less organelles (PMLOs), which are commonly found in eukaryotic cells and bacteria [4–12]. Since PMLOs reversibly and controllably isolate target molecules in specialized compartments, they constitute an intricate answer to the cellular need to facilitate and control molecular interactions [5]. In fact, PMLOs serve as an important complement to the common membrane-encapsulated organelles, such as nucleus, mitochondria, Golgi apparatus, Golgi vesicles, smooth endoplasmic reticulum, rough endoplasmic reticulum, lysosomes, peroxisomes, secretory vesicles or granules (e.g., insulin granules), chloroplasts, and vacuoles. These membrane-bound organelles represent evolutionarily conserved compartments with complex barriers (membranes) permitting spatial isolation as well as energy-efficient and passive buffering of stochastic events [13].

Although traditional membrane-encapsulated organelles represent functionally optimized (and evolutionary conserved) compartments, where membranes provide the physical separation within a cell needed for some specialized processes to occur, PMLOs, which are also functionally optimized compartments, are not surrounded by a membrane (as follows from their name). PMLOs represent condensed heterogeneous liquid-like mixtures of proteins and nucleic acids formed via liquid–liquid phase separation (LLPS) or biological liquid–liquid phase transitions (LLPTs).

By concentrating specific proteins (and frequently RNA and/or DNA), biological LLPTs generate PMLOs, which are considered as intracellular functional hot spots that serve as organizers of cellular biochemistry [14,15]. The resulting PMLOs are many, and cytoplasmic granules include centrosomes [16], germline P-granules (germ cell granules or nuage) [17,18], neuronal RNA granules [19], processing bodies or P-bodies [20], and stress granules (SGs) [21]. There is only one type of PMLO in mitochondria and in chloroplasts, chloroplast SGs and mitochondrial RNA granules. On the other hand, the nucleus contains a large realm of nuclear PMLOs, such as nucleoli [22], nuclear pores [23], chromatin [24], Cajal bodies (CBs; [25]), nuclear stress bodies (nSBs) [26,27], nuclear gems (Gemini of coiled bodies or Gemini of Cajal bodies) [28,29], Sam68 nuclear bodies (SNBs) [30], perinucleolar compartment (PNC) [30], promyelocytic leukemia nuclear bodies (PML nuclear bodies) or PML oncogenic domains (PODs) [31], PcG bodies (polycomb bodies, subnuclear organelles containing polycomb group proteins) [32], paraspeckles [33], Oct1/PTF/transcription (OPT) domains [34], nuclear speckles or interchromatin granule clusters [35], histone locus bodies (HLBs) [36], and cleavage bodies [37], to name a few. This list represents only the tip of the iceberg, as recent studies suggest that 50+ different PMLOs can be found in eukaryotic cells and bacteria [4,5], and this number is increasing on a regular basis.

PMLOs are characterized by different physical properties, molecular compositions, subcellular localizations, cell type-specific features, and fast responsiveness to changes in cellular surroundings and environmental cues. In fact, PMLOs are dynamic, cell size-dependent, liquid-like bodies [9] with dimensions ranging from tens of nm to tens of μm and specific cellular distributions [11]. On the other hand, it has been shown that, although many intracellular bodies are liquid-like droplets with highly dynamic organization [10,15,38–43], some other PMLOs, e.g., amyloid bodies, centrosomes, nuclear pores, and Balbiani bodies, are characterized as "bioreactive gels" whose properties vary from solid-like to gels and viscous liquids [44]. Also, PMLOs are characterized by a high variability of their organizational complexity and compositions. In fact, based on their protein compositions (number of droplet-specific proteins), human PMLOs can be arranged in the following order: nucleolus (1626) > chromatin (1350) > nuclear speckles (650) > centrosome (530) > mitochondrial RNA granules (229) > promyelocytic leukemia protein (PML) nuclear bodies (104) > SGs (57) > perinuclear compartment (55) > Cajal bodies (CBs) (54) > polycomb group (PcG) bodies (48) > P-granules (Perinuclear RNA granules specific to the germline) (19) > nuage (18) > cleavage bodies (14) > Gemini (10) > SAM68 bodies (8) > paraspeckles (6) > nuclear SGs (5) = OPT (Oct1/PTF/transcription) domain (5) > histone locus bodies (HLBs) (4) = neuronal ribonucleoprotein (RNP) granules (4) [45]. Furthermore, environmental changes can also affect the protein composition and the physical properties of PMLOs [11], and this variability is controlled by different cellular factors, including (but not limited to) the stage of the cell cycle, the presence of growth stimuli, or stress [11].

LLPTs causing the PMLO formation may be triggered by a variety of environmental factors, such as: fluctuations in levels of biomacromolecules (proteins and nucleic acids) undergoing phase separation; variations in the concentrations of specific small molecules or salts; changes in temperature, osmolarity, and/or pH of the solution; various alterations of the phase-forming proteins caused by a multitude of posttranslational modifications (PTMs), alternative splicing, or binding of certain partners; or alterations of the environmental conditions modulating the protein–nucleic acid or the protein–protein interactions [8,9,14,46,47]. One should also keep in mind that the biological LLPTs and the related processes of PMLO formation are strongly condition-dependent, completely reversible, and tightly controlled [4,5]. This is schematically represented by Figure 1 showing LLPT and factors triggering these transitions.

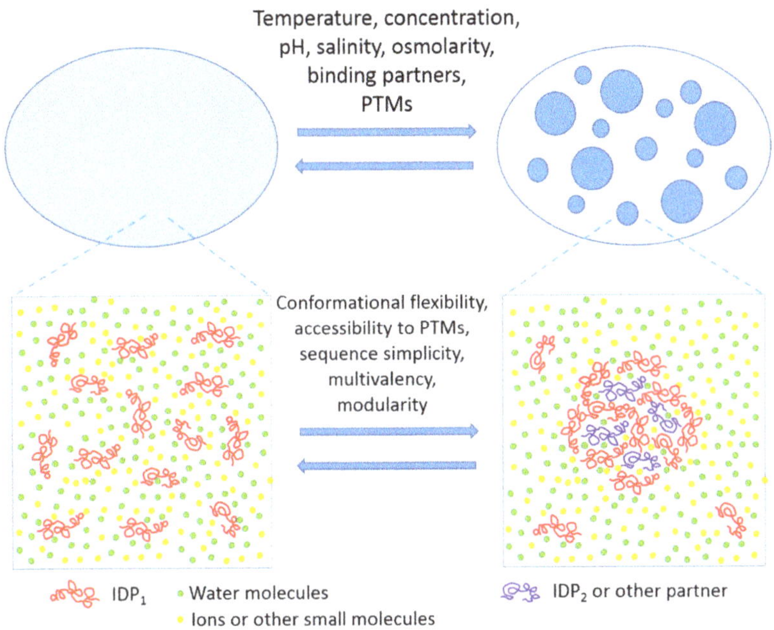

Figure 1. Schematic representation of liquid–liquid phase transitions (LLPT) and thermodynamic factors (top) and intrinsic disorder-related features controlling liquid–liquid phase transitions in protein solutions. This figure is reprinted from Current Opinion in Structural Biology, Vol. 44, Uversky V.N. Intrinsically disordered proteins in overcrowded milieu: membrane-less organelles, phase separation, and intrinsic disorder, Pages No. 18–30, Copyright 2017, with permission from Elsevier.

Obviously, since there are no membranes around PMLOs, their biogenesis and structural coherence are exclusively governed by the intra-organelle protein–protein, protein–RNA, and/or protein–DNA interactions [48]. Furthermore, due to lack of surrounding membranes, the components of PMLOs are not protected from the environment and rapidly circulate between the organelle and its adjoining surroundings [49,50]. As a result, PMLOs exhibit several features characteristic of liquids. In fact, they show wettability (i.e., they can uphold contact with a solid surface) and possess sufficient surface tension for maintenance of their spherical shape. They can fuse upon contact, flow in response to shear stresses, and drip [17,21,51,52]. Therefore, based on their properties, PMLOs can be classified as a special liquid state (or liquid phase) of cytoplasm, matrix, nucleoplasm, or stroma characterized by the major physico-chemical properties that are rather close to the features of the corresponding intracellular fluids in which they are found [9]. On the other hand, although the intrinsic density and the viscosity of many PMLOs are relatively low, being not very different from those of the cytoplasm or the nucleoplasm [17,21,51–56], the PMLO interior is classified as an overcrowded milieu [4]. This is due to the fact that PMLOs contain noticeably higher total protein concentrations than those found within the crowded cytoplasm and the nucleoplasm [4]. An illustrative example of this overcrowded nature of PMLOs is given by nucleoli, speckles, and Cajal bodies of the *Xenopus* oocyte nucleus with the total protein concentrations of 215, 162, and 136 mg/mL, respectively. These values are noticeably higher than the total protein concentration of 106 mg/mL in the surrounding nucleoplasm [55]. More globally, although the dilute phase in a cell is maintained at the critical phase separation concentrations of proteins and nucleic acids [11], these biomacromolecules can be concentrated ~10–100-fold within the droplets [53,57], reaching millimolar concentrations [58].

Importantly, recent studies revealed that PMLOs are not homogeneous themselves. In fact, SGs were shown to be characterized by a heterogeneous structure, where the core was more densely packed and less easily accessible than the more diffused shell with easier exchange of the constituents between the SGs and the adjacent cytoplasm [59]. Because the components of a dense core are brought together at early stages of the SG assembly, whereas a diffused shell of these PMLOs is formed at later steps, these different SG phases are kinetically formed at different stages of the SG assembly [59]. Furthermore, using a combination of various in vivo and in vitro approaches with computational modeling, it was recently shown that one of the most studied PMLOs, the nucleolus, possesses layered droplet organization containing internal sub-compartments [60]. These sub-compartments were shown to represent distinct, coexisting, non-coalescing liquid phases formed by LLPTs of specific nucleolar proteins, suggesting that biological phase separation can generate multilayered liquids [60].

PMLOs are crucial for cellular functionality and are now considered as key organizers and regulators of many cellular processes [11]. Since multiple cellular components are concentrated within the PMLOs, they regulate a broad cohort of cellular processes ranging from transcription to translational repression, to RNP assembly and processing, to biogenesis of ribosomes, to transport and degradation of mRNA, and to intracellular signaling [15]. Because the LLPTs causing PMLOs are fast under normal physiological conditions, and because the PMLO components are concentrated in a dynamic, selective, and reversible manner, such intracellular bodies are well suited for processing of local information and for handling rapid and controllable responses to environmental alterations, indicating that at least some PMLOs can serve as dynamic sensors of localized signals [61].

Normally, the highly dynamic structure and composition of polyfunctional and multicomponent PMLOs allow them to provide finely tuned regulation of various intracellular processes. On the other hand, as with many other protein intrinsic disorder-based events and activities [62], even the slightest disruption of the activity of PMLOs and their biogenesis can lead to an imbalance of intracellular regulatory pathways, resulting in the development of various pathological conditions [40,63–73]. For example, although in their normal state, the majority of PMLOs (including SGs) possess liquid-like properties, their aging can promote development of a much less dynamic state that typically coincides with the appearance of fibrous structures [74]. Such aging-related alterations in the mechanical and the physical properties of PMLOs can be of biological and pathological significance [74]. For example, it was shown that the time-dependent changes in the dense core of aging SGs can promote formation of insoluble protein aggregates linked to neurodegenerative diseases [70,75].

2. Proteinaceous Membrane-Less Organelles, Liquid–liquid Phase Transitions, and Intrinsic Disorder

The facts presented in the previous section indicate that, typically, specific sets of resident proteins can be found in PMLOs. Among the characteristic properties uniting many of these PMLO-residing proteins is the presence of high intrinsic disorder levels, suggesting the overall importance of intrinsically disordered proteins (IDPs) or hybrid proteins with ordered domains and intrinsically disordered protein regions (IDPRs) for LLPTs and PMLOs [4,5,8,45,74,76–84]. In fact, the biogenesis of several PMLOs (e.g., nuages [57], nucleolus [85], P-granules [80], and RNA granules [74]) was shown to be critically dependent on IDPs/IDPRs. This is because the LLPTs driving the PMLO formation are determined by weak multivalent interactions between multi-domain proteins and/or IDPs, hybrid proteins with ordered domains and IDPRs [4,5,86], proteins with RNA-binding domains [87], proteins containing repeats of amino acids with polar and charged groups, or proteins with low complexity domains (LCDs) [5,9,88].

There are multiple reasons for why IDPs/IDPRs serve as the most appropriate candidates for biological LLPTs leading to PMLO formation. These reasons include: the overall high abundance of IDPs/IDPRs in various proteomes [89–93] {e.g., among the eukaryotic proteins, ~25–30% are mostly disordered [91], long IDPRs (longer than 30 residues) are found in more than half of eukaryotic proteins [89–91], whereas such long IDPRs are present in >70% of signaling proteins [94]}; their lack of fixed structure [95–100]; their high spatio-temporal heterogeneity and mosaic structural organization that constitute a mix of foldons, inducible foldons, morphing inducible foldons, non-foldons, semi-foldons, and unfoldons [86,100–103]; the ability of these proteins to serve as highly promiscuous binders engaged in a multitude of interactions with highly diversified partners and to thereby regulate and control a wide spectrum of cellular processes [95,97–100,104–108]; and their ability to preserve their mostly disordered status within PMLOs that defines the fluidity of these organelles and determines PMLOs as supramolecular fuzzy complexes (see below).

Weak, multivalent, and rather non-specific interactions between one or more IDPs/IDPRs and between IDPs/IDPRs and nucleic acids are expected to drive biological LLPTs, leading to the PMLO formation. The physico-chemical nature of these interactions driving phase separation can be highly diversified and range from π–π contacts to cation–π interactions [15], to hydrophobic interactions, and to heterologous and homologous electrostatic attraction between differently charged biological polymers and differently charged parts of the same protein molecules [4,5]. By virtue of the peculiarities of their amino acid sequences and biophysical properties, IDPs/IDPRs are uniquely positioned in the category of biological macromolecules capable of undergoing LLPTs and controlling the biogenesis of PMLOs. For example, the conformational behavior of IDPs/IDPRs is, at least in part, determined by the presence of a large number of charged residues and depletion in hydrophobic residues [95], which explains the mostly electrostatic nature of their interactions [109]. Since IDPs/IDPRs do not possess stable structures, existing in a form of highly dynamic conformational ensembles of rapidly interconverting flexible structures, mean electrostatic fields are created that are used in polyelectrostatic attraction [110]. Furthermore, since charged residues are typically heterogeneously distributed within the amino acid sequences of many IDPs/IDPRs, patches of similarly charged residues are generated, and such "block co-polymer"-like structure might serve as a good template for the electrostatics-driven LLPTs [5]. More generally, common presence in IDPs/IDPRs of arrays of tandem repeats with different physico-chemical properties [111] creates a foundation of the flexible multivalency needed for LLPTs [5]. Also, IDPs/IDPRs are known to be commonly subjected to various post-translational modifications (PTMs) [112,113]. As LLPTs can be regulated by PTMs [53], this PTM-controlled conformational and functional variability of IDPs/IDPRs is very appropriate for the regulation of PMLO biogenesis [5]. Being the "edge of chaos" systems [63,101,114,115], IDPs/IDPRs are known for their high sensitivity and responsiveness to (even rather subtle) environmental changes. Because of this environmental sensitivity and receptivity as well as the capability to undergo fast, highly controllable, environment-modulated transitions, IDPs/IDPRs play crucial roles in the regulation of LLPTs and PMLOs [5].

3. Dysregulation of the Biogenesis of Intracellular Liquid Droplets and Disease

It was pointed out that, since the local concentrations of proteins in PMLOs are noticeably higher than those in the surrounding crowded media (and, as a result, the interior of PMLOs is considered as the overcrowded milieu [4]), and since some amyloidogenic proteins can be found in PMLOs and many of these proteins can undergo LLPS both in vitro and in vivo, dysregulation of the biogenesis of intracellular liquid droplets can be related to various human diseases [63]. This suggests the existence of a definite spatio-temporal window of safe existence, where a given PMLO appears at a definite cell location in a response to a definite environmental cue and exists there for a definite amount of time, whereas the pathological conversion from liquid to solid or gel form within the highly concentrated milieu of PMLO might happen outside of this window of safe existence [40,63,64,67,69–72,74,75,79,116–118]. Generally, molecular mechanisms associated with the said pathological transformations are related to the dysregulated biogenesis of PMLOs, eventually leading to the distortions of their dynamics and the promotion of pathological aggregation. Some of these mechanisms include pathological "aging" of PMLOs (or going beyond the safe time window), increased content of proteins involved in LLPTs, aberrant PTMs, some chromosomal translocation, and pathological mutations [73]. Among proteins for which the aberrant LLPTs are associated with pathological aggregation are TAR DNA binding protein-43 (TDP-43) linked to amyotrophic lateral sclerosis (ALS) [116], microtubule-associated protein tau involved in Alzheimer's disease (AD) [117,118], α-synuclein associated with Parkinson's disease (PD) [66], TDP-43, heterogeneous nuclear ribonucleoprotein A1 (hnRNPA1) linked to ALS [66], fused in sarcoma (FUS) associated with the pathogenesis of ALS and frontotemporal lobar degeneration (FTLD) [119], prion protein [120], and many RNA-binding IDPs possessing low complexity domains (LCD) [64]. In other words, pathogenic transformations of PMLOs are often associated with the decreased fuzziness of these intracellular liquid droplets.

4. Supramolecular Fuzziness of Intracellular Liquid Droplets

An important feature of PMLOs is their fluidity. The liquid-like properties of phase-separated droplets facilitate the functions of their constituents, which are accumulated within droplets at high concentrations and show slowed diffusion but remain dynamic. In fact, the concentrations of proteins residing within these liquid droplets can be ~10–100-higher than the protein content of the dilute phase [53,57]. Furthermore, being intrinsically disordered, these PMLO-residing proteins can be engaged in multivalent interactions. These observations raise an important question regarding how fluidity can be preserved within the overcrowded milieu of the PMLO interior. It is likely that one can find an answer to this question by analyzing the structural properties of proteins within PMLOs or artificial phase-separated liquid droplets. In fact, if an IDP/IDPR would undergo global folding as a result of an LLPT, then the resulting condensed phase would not be liquid but would contain ordered protein–protein or protein–RNA complexes stabilized by multivalent rigid body–rigid body interactions. Therefore, the fact that PMLOs are liquid indicates that IDPs/IDPRs undergoing LLPTs preserve high levels of intrinsic disorder. Several recent NMR-based studies are in agreement with this hypothesis [121]. For example, the DEAD box protein 4 (DDX4), which is a probable ATP-dependent RNA helicase that serves as a primary constituent of nuage or germ granules [122], was shown by ^{1}H-^{15}N HSQC spectroscopy to remain disordered within the droplets [57]. Similarly, the LCD of the fused in sarcoma (FUS) protein, which is associated with two devastating neurodegenerative disorders—amyotrophic lateral sclerosis (ALS) and frontotemporal dementia (FTD) [123]—remained mostly disordered within the droplet phase [58]. Also, the microtubule associated protein tau, which is an IDP involved in Alzheimer's disease [124,125] and other tauopathies [126], was shown to undergo LLPT in solution in a phosphorylation-dependent manner and preserved disordered state in the condensed phase [117]. Another IDP, BUB3-interacting and GLEBS motif-containing protein (BuGZ), the phase separation of which is involved in spindle matrix formation and function [127], was shown to remain dynamic in the spindle and its matrix [128].

The capability of IDPs/IDPRs to preserve high levels of disorder in their bound states is known as fuzziness, an important phenomenon emphasizing that formation, function, and/or regulation of the protein-based complexes/assembles are critically dependent on the intrinsic disorder of the constituent proteins [129,130]. Furthermore, it was emphasized that the biological activity of the resulting fuzzy complexes could be affected by fuzzy regions, which not only remained disordered but often escalated their conformational diversity in the bound state [131]. In fact, fuzzy regions are engaged in transient interactions, thereby establishing alternate contacts with specific partners. Flexibility and interactability of such regions can be regulated and controlled by PTMs and alternative splicing [131]. Because of the preservation of high disorder levels, PMLOs and artificial phase-separated liquid droplets represent fuzzy supramolecular complexes.

5. Conclusions

In summary, data accumulated to date indicate that high levels of intrinsic disorder are found in many PMLO resident proteins and show that the PMLO formation often relies on IDPs/IDPRs, indicating that PMLO biogenesis is crucially dependent on intrinsic disorder [8]. In other words, the lack of stable structure in IDPs/IDPRs, the ability of such proteins to be engaged in highly dynamic, weak, multivalent interactions combined with their capability to retain a highly mobile character after undergoing LLPTs define the liquid-like nature of PMLOs [5]. It is likely that the structural resilience of PMLOs and their capability to exist as stable entities in the absence of enclosing membranes combined with the free exchange of the constituents with the environment are also defined by the same properties of IDPs/IDPRs [5]. In summary, PMLOs are an enthralling form of disorder-based protein assemblages [4,5,86], which are formed without noticeable structural changes or ordering of their constituent IDPs/IDPRs when undergoing LLPTs, and which, as a result, are characterized by a highly dynamic nature defining their liquid-like appearance [57]. In other words, supramolecular fuzziness is crucial for many aspects of PMLO biogenesis, stability, and functionality.

Author Contributions: Conceptualization, validation, formal analysis, investigation, writing, V.N.U.

Funding: This research received no external funding.

Conflicts of Interest: The author declares no conflict of interest.

References

1. Zimmerman, S.B.; Trach, S.O. Estimation of macromolecule concentrations and excluded volume effects for the cytoplasm of escherichia coli. *J. Mol. Biol.* **1991**, *222*, 599–620. [CrossRef]
2. Rivas, G.; Ferrone, F.; Herzfeld, J. Life in a crowded world. *EMBO Rep.* **2004**, *5*, 23–27. [CrossRef] [PubMed]
3. Milo, R.; Jorgensen, P.; Moran, U.; Weber, G.; Springer, M. Bionumbers–the database of key numbers in molecular and cell biology. *Nucleic Acids Res.* **2010**, *38*, D750–D753. [CrossRef] [PubMed]
4. Uversky, V.N. Intrinsically disordered proteins in overcrowded milieu: Membrane-less organelles, phase separation, and intrinsic disorder. *Curr. Opin. Struct. Biol.* **2017**, *44*, 18–30. [CrossRef] [PubMed]
5. Uversky, V.N. Protein intrinsic disorder-based liquid-liquid phase transitions in biological systems: Complex coacervates and membrane-less organelles. *Adv. Colloid Interface Sci.* **2017**, *239*, 97–114. [CrossRef] [PubMed]
6. Zaslavsky, B.Y.; Uversky, V.N. In aqua veritas: The indispensable yet mostly ignored role of water in phase separation and membrane-less organelles. *Biochemistry* **2018**, *57*, 2437–2451. [CrossRef] [PubMed]
7. Zaslavsky, B.Y.; Ferreira, L.A.; Darling, A.L.; Uversky, V.N. The solvent side of proteinaceous membrane-less organelles in light of aqueous two-phase systems. *Int. J. Biol. Macromol.* **2018**, *117*, 1224–1251. [CrossRef] [PubMed]
8. Uversky, V.N.; Kuznetsova, I.M.; Turoverov, K.K.; Zaslavsky, B. Intrinsically disordered proteins as crucial constituents of cellular aqueous two phase systems and coacervates. *FEBS Lett.* **2015**, *589*, 15–22. [CrossRef]
9. Brangwynne, C.P. Phase transitions and size scaling of membrane-less organelles. *J. Cell Biol.* **2013**, *203*, 875–881. [CrossRef]
10. Hyman, A.A.; Weber, C.A.; Julicher, F. Liquid-liquid phase separation in biology. *Annu. Rev. Cell Dev. Biol.* **2014**, *30*, 39–58. [CrossRef]
11. Mitrea, D.M.; Kriwacki, R.W. Phase separation in biology; functional organization of a higher order. *Cell Commun. Signal.* **2016**, *14*, 1. [CrossRef] [PubMed]
12. Gomes, E.; Shorter, J. The molecular language of membraneless organelles. *J. Biol. Chem.* **2019**, *294*, 7115–7127. [CrossRef] [PubMed]
13. Stoeger, T.; Battich, N.; Pelkmans, L. Passive noise filtering by cellular compartmentalization. *Cell* **2016**, *164*, 1151–1161. [CrossRef] [PubMed]
14. Zhu, L.; Brangwynne, C.P. Nuclear bodies: The emerging biophysics of nucleoplasmic phases. *Curr. Opin. Cell Biol.* **2015**, *34*, 23–30. [CrossRef] [PubMed]
15. Banani, S.F.; Lee, H.O.; Hyman, A.A.; Rosen, M.K. Biomolecular condensates: Organizers of cellular biochemistry. *Nat. Rev. Mol. Cell Biol.* **2017**, *18*, 285–298. [CrossRef] [PubMed]

16. Decker, M.; Jaensch, S.; Pozniakovsky, A.; Zinke, A.; O'Connell, K.F.; Zachariae, W.; Myers, E.; Hyman, A.A. Limiting amounts of centrosome material set centrosome size in c. Elegans embryos. *Curr. Biol: CB* **2011**, *21*, 1259–1267. [CrossRef]
17. Brangwynne, C.P.; Eckmann, C.R.; Courson, D.S.; Rybarska, A.; Hoege, C.; Gharakhani, J.; Julicher, F.; Hyman, A.A. Germline p granules are liquid droplets that localize by controlled dissolution/condensation. *Science* **2009**, *324*, 1729–1732. [CrossRef]
18. Chuma, S.; Hosokawa, M.; Tanaka, T.; Nakatsuji, N. Ultrastructural characterization of spermatogenesis and its evolutionary conservation in the germline: Germinal granules in mammals. *Mol. Cell. Endocrinol.* **2009**, *306*, 17–23. [CrossRef]
19. Kiebler, M.A.; Bassell, G.J. Neuronal rna granules: Movers and makers. *Neuron* **2006**, *51*, 685–690. [CrossRef]
20. Decker, C.J.; Teixeira, D.; Parker, R. Edc3p and a glutamine/asparagine-rich domain of lsm4p function in processing body assembly in saccharomyces cerevisiae. *J. Cell Biol.* **2007**, *179*, 437–449. [CrossRef]
21. Wippich, F.; Bodenmiller, B.; Trajkovska, M.G.; Wanka, S.; Aebersold, R.; Pelkmans, L. Dual specificity kinase dyrk3 couples stress granule condensation/dissolution to mtorc1 signaling. *Cell* **2013**, *152*, 791–805. [CrossRef] [PubMed]
22. Shav-Tal, Y.; Blechman, J.; Darzacq, X.; Montagna, C.; Dye, B.T.; Patton, J.G.; Singer, R.H.; Zipori, D. Dynamic sorting of nuclear components into distinct nucleolar caps during transcriptional inhibition. *Mol. Biol. Cell* **2005**, *16*, 2395–2413. [CrossRef] [PubMed]
23. Grossman, E.; Medalia, O.; Zwerger, M. Functional architecture of the nuclear pore complex. *Annu. Rev. Biophys.* **2012**, *41*, 557–584. [CrossRef] [PubMed]
24. Li, B.; Carey, M.; Workman, J.L. The role of chromatin during transcription. *Cell* **2007**, *128*, 707–719. [CrossRef] [PubMed]
25. Strzelecka, M.; Trowitzsch, S.; Weber, G.; Luhrmann, R.; Oates, A.C.; Neugebauer, K.M. Coilin-dependent snrnp assembly is essential for zebrafish embryogenesis. *Nat. Struct. Mol. Biol.* **2010**, *17*, 403–409. [CrossRef] [PubMed]
26. Biamonti, G.; Vourc'h, C. Nuclear stress bodies. *Cold Spring Harb. Perspect. Biol.* **2010**, *2*, a000695. [CrossRef]
27. Biamonti, G. Nuclear stress bodies: A heterochromatin affair? *Nat. Rev. Mol. Cell Biol.* **2004**, *5*, 493–498. [CrossRef] [PubMed]
28. Matera, A.G.; Frey, M.R. Coiled bodies and gems: Janus or gemini? *Am. J. Hum. Genet.* **1998**, *63*, 317–321. [CrossRef]
29. Gubitz, A.K.; Feng, W.; Dreyfuss, G. The smn complex. *Exp. Cell Res.* **2004**, *296*, 51–56. [CrossRef]
30. Huang, S. Review: Perinucleolar structures. *J. Struct. Biol.* **2000**, *129*, 233–240. [CrossRef]
31. Maul, G.G.; Negorev, D.; Bell, P.; Ishov, A.M. Review: Properties and assembly mechanisms of nd10, pml bodies, or pods. *J. Struct. Biol.* **2000**, *129*, 278–287. [CrossRef] [PubMed]
32. Pirrotta, V.; Li, H.B. A view of nuclear polycomb bodies. *Curr. Opin. Genet. Dev.* **2012**, *22*, 101–109. [CrossRef] [PubMed]
33. Fox, A.H.; Lam, Y.W.; Leung, A.K.; Lyon, C.E.; Andersen, J.; Mann, M.; Lamond, A.I. Paraspeckles: A novel nuclear domain. *Curr. Biol.: CB* **2002**, *12*, 13–25. [CrossRef]
34. Harrigan, J.A.; Belotserkovskaya, R.; Coates, J.; Dimitrova, D.S.; Polo, S.E.; Bradshaw, C.R.; Fraser, P.; Jackson, S.P. Replication stress induces 53bp1-containing opt domains in g1 cells. *J. Cell Biol.* **2011**, *193*, 97–108. [CrossRef] [PubMed]
35. Lamond, A.I.; Spector, D.L. Nuclear speckles: A model for nuclear organelles. *Nat. Rev. Mol. Cell Biol.* **2003**, *4*, 605–612. [CrossRef] [PubMed]
36. Nizami, Z.; Deryusheva, S.; Gall, J.G. The cajal body and histone locus body. *Cold Spring Harb. Perspect. Biol.* **2010**, *2*, a000653. [CrossRef] [PubMed]
37. Li, L.; Roy, K.; Katyal, S.; Sun, X.; Bleoo, S.; Godbout, R. Dynamic nature of cleavage bodies and their spatial relationship to ddx1 bodies, cajal bodies, and gems. *Mol. Biol. Cell* **2006**, *17*, 1126–1140. [CrossRef]
38. Alberti, S. The wisdom of crowds: Regulating cell function through condensed states of living matter. *J. Cell. Sci.* **2017**, *130*, 2789–2796. [CrossRef]
39. Courchaine, E.M.; Lu, A.; Neugebauer, K.M. Droplet organelles? *EMBO J.* **2016**, *35*, 1603–1612. [CrossRef]
40. Shin, Y.; Brangwynne, C.P. Liquid phase condensation in cell physiology and disease. *Science* **2017**, *357*, eaaf4382. [CrossRef]

41. Chong, P.A.; Forman-Kay, J.D. Liquid-liquid phase separation in cellular signaling systems. *Curr. Opin. Struct. Biol.* **2016**, *41*, 180–186. [CrossRef] [PubMed]
42. Hnisz, D.; Shrinivas, K.; Young, R.A.; Chakraborty, A.K.; Sharp, P.A. A phase separation model for transcriptional control. *Cell* **2017**, *169*, 13–23. [CrossRef] [PubMed]
43. Saha, S.; Weber, C.A.; Nousch, M.; Adame-Arana, O.; Hoege, C.; Hein, M.Y.; Osborne-Nishimura, E.; Mahamid, J.; Jahnel, M.; Jawerth, L.; et al. Polar positioning of phase-separated liquid compartments in cells regulated by an mrna competition mechanism. *Cell* **2016**, *166*, 1572–1584. [CrossRef] [PubMed]
44. Woodruff, J.B.; Hyman, A.A.; Boke, E. Organization and function of non-dynamic biomolecular condensates. *Trends Biochem. Sci.* **2018**, *43*, 81–94. [CrossRef]
45. Darling, A.L.; Liu, Y.; Oldfield, C.J.; Uversky, V.N. Intrinsically disordered proteome of human membrane-less organelles. *Proteomics* **2018**, *18*, e1700193. [CrossRef]
46. Brangwynne, C.P.; Tompa, P.; Pappu, R.V. Polymer physics of intracellular phase transitions. *Nat. Phys.* **2015**, *11*, 899–904. [CrossRef]
47. Holehouse, A.S.; Pappu, R.V. Functional implications of intracellular phase transitions. *Biochemistry* **2018**. [CrossRef]
48. Dundr, M.; Misteli, T. Biogenesis of nuclear bodies. *Cold Spring Harb. Perspect. Biol.* **2010**, *2*, a000711. [CrossRef]
49. Phair, R.D.; Misteli, T. High mobility of proteins in the mammalian cell nucleus. *Nature* **2000**, *404*, 604–609. [CrossRef]
50. Pederson, T. Protein mobility within the nucleus–what are the right moves? *Cell* **2001**, *104*, 635–638. [CrossRef]
51. Brangwynne, C.P.; Mitchison, T.J.; Hyman, A.A. Active liquid-like behavior of nucleoli determines their size and shape in xenopus laevis oocytes. *Proc. Natl. Acad. Sci. United States Am.* **2011**, *108*, 4334–4339. [CrossRef] [PubMed]
52. Feric, M.; Brangwynne, C.P. A nuclear f-actin scaffold stabilizes ribonucleoprotein droplets against gravity in large cells. *Nat. Cell Biol.* **2013**, *15*, 1253–1259. [CrossRef] [PubMed]
53. Li, P.; Banjade, S.; Cheng, H.C.; Kim, S.; Chen, B.; Guo, L.; Llaguno, M.; Hollingsworth, J.V.; King, D.S.; Banani, S.F.; et al. Phase transitions in the assembly of multivalent signalling proteins. *Nature* **2012**, *483*, 336–340. [CrossRef] [PubMed]
54. Aggarwal, S.; Snaidero, N.; Pahler, G.; Frey, S.; Sanchez, P.; Zweckstetter, M.; Janshoff, A.; Schneider, A.; Weil, M.T.; Schaap, I.A.; et al. Myelin membrane assembly is driven by a phase transition of myelin basic proteins into a cohesive protein meshwork. *PLoS Biol.* **2013**, *11*, e1001577. [CrossRef] [PubMed]
55. Handwerger, K.E.; Cordero, J.A.; Gall, J.G. Cajal bodies, nucleoli, and speckles in the xenopus oocyte nucleus have a low-density, sponge-like structure. *Mol. Biol. Cell* **2005**, *16*, 202–211. [CrossRef] [PubMed]
56. Updike, D.L.; Hachey, S.J.; Kreher, J.; Strome, S. P granules extend the nuclear pore complex environment in the c. Elegans germ line. *J. Cell Biol.* **2011**, *192*, 939–948. [CrossRef] [PubMed]
57. Nott, T.J.; Petsalaki, E.; Farber, P.; Jervis, D.; Fussner, E.; Plochowietz, A.; Craggs, T.D.; Bazett-Jones, D.P.; Pawson, T.; Forman-Kay, J.D.; et al. Phase transition of a disordered nuage protein generates environmentally responsive membraneless organelles. *Mol. Cell* **2015**, *57*, 936–947. [CrossRef] [PubMed]
58. Burke, K.A.; Janke, A.M.; Rhine, C.L.; Fawzi, N.L. Residue-by-residue view of in vitro fus granules that bind the c-terminal domain of rna polymerase ii. *Mol. Cell* **2015**, *60*, 231–241. [CrossRef] [PubMed]
59. Wheeler, J.R.; Matheny, T.; Jain, S.; Abrisch, R.; Parker, R. Distinct stages in stress granule assembly and disassembly. *eLife* **2016**, *5*. [CrossRef] [PubMed]
60. Feric, M.; Vaidya, N.; Harmon, T.S.; Mitrea, D.M.; Zhu, L.; Richardson, T.M.; Kriwacki, R.W.; Pappu, R.V.; Brangwynne, C.P. Coexisting liquid phases underlie nucleolar subcompartments. *Cell* **2016**, *165*, 1686–1697. [CrossRef] [PubMed]
61. Formicola, N.; Vijayakumar, J.; Besse, F. Neuronal rnp granules: Dynamic sensors of localized signals. *Traffic* **2019**, *20*, 639–649. [PubMed]
62. Uversky, V.N. Wrecked regulation of intrinsically disordered proteins in diseases: Pathogenicity of deregulated regulators. *Front. Mol. Biosci* **2014**, *1*, 6. [CrossRef] [PubMed]
63. Darling, A.L.; Zaslavsky, B.Y.; Uversky, V.N. Intrinsic disorder-based emergence in cellular biology: Physiological and pathological liquid-liquid phase transitions in cells. *Polymer* **2019**, *11*, 990. [CrossRef] [PubMed]
64. Aulas, A.; Vande Velde, C. Alterations in stress granule dynamics driven by tdp-43 and fus: A link to pathological inclusions in als? *Front. Cell Neurosci.* **2015**, *9*, 423. [CrossRef] [PubMed]

65. Lessnick, S.L.; Ladanyi, M. Molecular pathogenesis of ewing sarcoma: New therapeutic and transcriptional targets. *Annu. Rev. Pathol.* **2012**, *7*, 145–159. [CrossRef] [PubMed]
66. Liu, C.; Fang, Y. New insights of poly(adp-ribosylation) in neurodegenerative diseases: A focus on protein phase separation and pathologic aggregation. *Biochem. Pharm.* **2019**. [CrossRef]
67. Liu-Yesucevitz, L.; Bilgutay, A.; Zhang, Y.J.; Vanderweyde, T.; Citro, A.; Mehta, T.; Zaarur, N.; McKee, A.; Bowser, R.; Sherman, M.; et al. Tar DNA binding protein-43 (tdp-43) associates with stress granules: Analysis of cultured cells and pathological brain tissue. *PLoS ONE* **2010**, *5*, e13250. [CrossRef] [PubMed]
68. McGurk, L.; Gomes, E.; Guo, L.; Mojsilovic-Petrovic, J.; Tran, V.; Kalb, R.G.; Shorter, J.; Bonini, N.M. Poly(adp-ribose) prevents pathological phase separation of tdp-43 by promoting liquid demixing and stress granule localization. *Mol. Cell* **2018**, *71*, 703–717. [CrossRef]
69. Molliex, A.; Temirov, J.; Lee, J.; Coughlin, M.; Kanagaraj, A.P.; Kim, H.J.; Mittag, T.; Taylor, J.P. Phase separation by low complexity domains promotes stress granule assembly and drives pathological fibrillization. *Cell* **2015**, *163*, 123–133. [CrossRef]
70. Dobra, I.; Pankivskyi, S.; Samsonova, A.; Pastre, D.; Hamon, L. Relation between stress granules and cytoplasmic protein aggregates linked to neurodegenerative diseases. *Curr. Neurol. Neurosci. Rep.* **2018**, *18*, 107. [CrossRef]
71. Duan, Y.; Du, A.; Gu, J.; Duan, G.; Wang, C.; Gui, X.; Ma, Z.; Qian, B.; Deng, X.; Zhang, K.; et al. Parylation regulates stress granule dynamics, phase separation, and neurotoxicity of disease-related rna-binding proteins. *Cell Res.* **2019**, *29*, 233–247. [CrossRef]
72. Harrison, A.F.; Shorter, J. RNA-binding proteins with prion-like domains in health and disease. *Biochem. J.* **2017**, *474*, 1417–1438. [CrossRef] [PubMed]
73. Patel, A.; Lee, H.O.; Jawerth, L.; Maharana, S.; Jahnel, M.; Hein, M.Y.; Stoynov, S.; Mahamid, J.; Saha, S.; Franzmann, T.M.; et al. A liquid-to-solid phase transition of the als protein fus accelerated by disease mutation. *Cell* **2015**, *162*, 1066–1077. [CrossRef] [PubMed]
74. Lin, Y.; Protter, D.S.; Rosen, M.K.; Parker, R. Formation and maturation of phase-separated liquid droplets by rna-binding proteins. *Mol. Cell* **2015**, *60*, 208–219. [CrossRef] [PubMed]
75. Elbaum-Garfinkle, S. Matter over mind: Liquid phase separation and neurodegeneration. *J. Biol. Chem* **2019**, *294*, 7160–7168. [CrossRef] [PubMed]
76. Meng, F.; Na, I.; Kurgan, L.; Uversky, V.N. Compartmentalization and functionality of nuclear disorder: Intrinsic disorder and protein-protein interactions in intra-nuclear compartments. *Int. J. Mol. Sci.* **2015**, *17*. [CrossRef]
77. Toretsky, J.A.; Wright, P.E. Assemblages: Functional units formed by cellular phase separation. *J. Cell Biol.* **2014**, *206*, 579–588. [CrossRef] [PubMed]
78. Csizmok, V.; Follis, A.V.; Kriwacki, R.W.; Forman-Kay, J.D. Dynamic protein interaction networks and new structural paradigms in signaling. *Chem. Rev.* **2016**. [CrossRef]
79. Kedersha, N.; Ivanov, P.; Anderson, P. Stress granules and cell signaling: More than just a passing phase? *Trends Biochem. Sci* **2013**, *38*, 494–506. [CrossRef] [PubMed]
80. Elbaum-Garfinkle, S.; Kim, Y.; Szczepaniak, K.; Chen, C.C.; Eckmann, C.R.; Myong, S.; Brangwynne, C.P. The disordered p granule protein laf-1 drives phase separation into droplets with tunable viscosity and dynamics. *Proc. Natl. Acad. Sci. United States Am.* **2015**, *112*, 7189–7194. [CrossRef]
81. Pak, C.W.; Kosno, M.; Holehouse, A.S.; Padrick, S.B.; Mittal, A.; Ali, R.; Yunus, A.A.; Liu, D.R.; Pappu, R.V.; Rosen, M.K. Sequence determinants of intracellular phase separation by complex coacervation of a disordered protein. *Mol. Cell* **2016**, *63*, 72–85. [CrossRef] [PubMed]
82. Panas, M.D.; Ivanov, P.; Anderson, P. Mechanistic insights into mammalian stress granule dynamics. *J. Cell Biol.* **2016**, *215*, 313–323. [CrossRef] [PubMed]
83. Protter, D.S.; Parker, R. Principles and properties of stress granules. *Trends Cell Biol* **2016**, *26*, 668–679. [CrossRef] [PubMed]
84. Simon, J.R.; Carroll, N.J.; Rubinstein, M.; Chilkoti, A.; Lopez, G.P. Programming molecular self-assembly of intrinsically disordered proteins containing sequences of low complexity. *Nat. Chem.* **2017**, *9*, 509–515. [CrossRef] [PubMed]
85. Mitrea, D.M.; Cika, J.A.; Guy, C.S.; Ban, D.; Banerjee, P.R.; Stanley, C.B.; Nourse, A.; Deniz, A.A.; Kriwacki, R.W. Nucleophosmin integrates within the nucleolus via multi-modal interactions with proteins displaying r-rich linear motifs and rrna. *eLife* **2016**, *5*. [CrossRef]

86. Uversky, V.N. Functional roles of transiently and intrinsically disordered regions within proteins. *FEBS J.* **2015**, *282*, 1182–1189. [CrossRef]
87. Drino, A.; Schaefer, M.R. Rnas, phase separation, and membrane-less organelles: Are post-transcriptional modifications modulating organelle dynamics? *Bioessays* **2018**, *40*, e1800085. [CrossRef]
88. Toll-Riera, M.; Rado-Trilla, N.; Martys, F.; Alba, M.M. Role of low-complexity sequences in the formation of novel protein coding sequences. *Mol. Biol. Evol.* **2012**, *29*, 883–886. [CrossRef]
89. Dunker, A.K.; Obradovic, Z.; Romero, P.; Garner, E.C.; Brown, C.J. Intrinsic protein disorder in complete genomes. *Genome Inf. Ser. Workshop Genome Inf.* **2000**, *11*, 161–171.
90. Ward, J.J.; Sodhi, J.S.; McGuffin, L.J.; Buxton, B.F.; Jones, D.T. Prediction and functional analysis of native disorder in proteins from the three kingdoms of life. *J. Mol. Biol.* **2004**, *337*, 635–645. [CrossRef]
91. Oldfield, C.J.; Cheng, Y.; Cortese, M.S.; Brown, C.J.; Uversky, V.N.; Dunker, A.K. Comparing and combining predictors of mostly disordered proteins. *Biochemistry* **2005**, *44*, 1989–2000. [CrossRef] [PubMed]
92. Xue, B.; Dunker, A.K.; Uversky, V.N. Orderly order in protein intrinsic disorder distribution: Disorder in 3500 proteomes from viruses and the three domains of life. *J. Biomol. Struct. Dyn.* **2012**, *30*, 137–149. [CrossRef] [PubMed]
93. Peng, Z.; Yan, J.; Fan, X.; Mizianty, M.J.; Xue, B.; Wang, K.; Hu, G.; Uversky, V.N.; Kurgan, L. Exceptionally abundant exceptions: Comprehensive characterization of intrinsic disorder in all domains of life. *Cell. Mol. Life Sci.: CMLS* **2014**. [CrossRef] [PubMed]
94. Iakoucheva, L.M.; Brown, C.J.; Lawson, J.D.; Obradovic, Z.; Dunker, A.K. Intrinsic disorder in cell-signaling and cancer-associated proteins. *J. Mol. Biol.* **2002**, *323*, 573–584. [CrossRef]
95. Uversky, V.N.; Gillespie, J.R.; Fink, A.L. Why are "natively unfolded" proteins unstructured under physiologic conditions? *Proteins* **2000**, *41*, 415–427. [CrossRef]
96. Dunker, A.K.; Obradovic, Z. The protein trinity–linking function and disorder. *Nat. Biotechnol* **2001**, *19*, 805–806. [CrossRef] [PubMed]
97. Uversky, V.N. Natively unfolded proteins: A point where biology waits for physics. *Protein Sci* **2002**, *11*, 739–756. [CrossRef] [PubMed]
98. Uversky, V.N. What does it mean to be natively unfolded? *Eur J. Biochem* **2002**, *269*, 2–12. [CrossRef] [PubMed]
99. Uversky, V.N. Protein folding revisited. A polypeptide chain at the folding-misfolding-nonfolding cross-roads: Which way to go? *Cell. Mol. Life Sci.: CMLS* **2003**, *60*, 1852–1871. [CrossRef] [PubMed]
100. Uversky, V.N. The most important thing is the tail: Multitudinous functionalities of intrinsically disordered protein termini. *FEBS Lett* **2013**, *587*, 1891–1901. [CrossRef] [PubMed]
101. Uversky, V.N. Unusual biophysics of intrinsically disordered proteins. *Biochim. Et Biophys. Acta* **2013**, *1834*, 932–951. [CrossRef] [PubMed]
102. Uversky, V.N. Intrinsic disorder-based protein interactions and their modulators. *Curr. Pharm. Des.* **2013**, *19*, 4191–4213. [CrossRef] [PubMed]
103. Jakob, U.; Kriwacki, R.; Uversky, V.N. Conditionally and transiently disordered proteins: Awakening cryptic disorder to regulate protein function. *Chem. Rev.* **2014**, *114*, 6779–6805. [CrossRef]
104. Dunker, A.K.; Lawson, J.D.; Brown, C.J.; Williams, R.M.; Romero, P.; Oh, J.S.; Oldfield, C.J.; Campen, A.M.; Ratliff, C.M.; Hipps, K.W.; et al. Intrinsically disordered protein. *J. Mol. Graph. Model.* **2001**, *19*, 26–59. [CrossRef]
105. Habchi, J.; Tompa, P.; Longhi, S.; Uversky, V.N. Introducing protein intrinsic disorder. *Chem. Rev.* **2014**, *114*, 6561–6588. [CrossRef] [PubMed]
106. van der Lee, R.; Buljan, M.; Lang, B.; Weatheritt, R.J.; Daughdrill, G.W.; Dunker, A.K.; Fuxreiter, M.; Gough, J.; Gsponer, J.; Jones, D.T.; et al. Classification of intrinsically disordered regions and proteins. *Chem. Rev.* **2014**, *114*, 6589–6631. [CrossRef] [PubMed]
107. Dunker, A.K.; Brown, C.J.; Obradovic, Z. Identification and functions of usefully disordered proteins. *Adv. Protein Chem.* **2002**, *62*, 25–49. [PubMed]
108. Dunker, A.K.; Brown, C.J.; Lawson, J.D.; Iakoucheva, L.M.; Obradovic, Z. Intrinsic disorder and protein function. *Biochemistry* **2002**, *41*, 6573–6582. [CrossRef] [PubMed]
109. Radivojac, P.; Obradovic, Z.; Smith, D.K.; Zhu, G.; Vucetic, S.; Brown, C.J.; Lawson, J.D.; Dunker, A.K. Protein flexibility and intrinsic disorder. *Protein Sci.* **2004**, *13*, 71–80. [CrossRef]
110. Mittag, T.; Kay, L.E.; Forman-Kay, J.D. Protein dynamics and conformational disorder in molecular recognition. *J. Mol Recognit* **2010**, *23*, 105–116. [CrossRef]
111. Jorda, J.; Xue, B.; Uversky, V.N.; Kajava, A.V. Protein tandem repeat-the more perfect, the less structured. *FEBS J.* **2010**, *277*, 2673–2682. [CrossRef] [PubMed]

112. Iakoucheva, L.M.; Radivojac, P.; Brown, C.J.; O'Connor, T.R.; Sikes, J.G.; Obradovic, Z.; Dunker, A.K. The importance of intrinsic disorder for protein phosphorylation. *Nucleic Acids Res.* **2004**, *32*, 1037–1049. [CrossRef] [PubMed]
113. Pejaver, V.; Hsu, W.L.; Xin, F.; Dunker, A.K.; Uversky, V.N.; Radivojac, P. The structural and functional signatures of proteins that undergo multiple events of post-translational modification. *Protein Sci.* **2014**, *23*, 1077–1093. [CrossRef] [PubMed]
114. Turoverov, K.K.; Kuznetsova, I.M.; Fonin, A.V.; Darling, A.L.; Zaslavsky, B.Y.; Uversky, V.N. Stochasticity of biological soft matter: Emerging concepts in intrinsically disordered proteins and biological phase separation. *Trends Biochem. Sci.* **2019**, *44*, 716–728. [CrossRef] [PubMed]
115. Uversky, V.N. Intrinsically disordered proteins and their "mysterious" (meta)physics. *Front. Phys.* **2019**, *7*. [CrossRef]
116. Babinchak, W.M.; Haider, R.; Dumm, B.K.; Sarkar, P.; Surewicz, K.; Choi, J.K.; Surewicz, W.K. The role of liquid-liquid phase separation in aggregation of the tdp-43 low-complexity domain. *J. Biol. Chem.* **2019**, *294*, 6306–6317. [CrossRef]
117. Ambadipudi, S.; Biernat, J.; Riedel, D.; Mandelkow, E.; Zweckstetter, M. Liquid-liquid phase separation of the microtubule-binding repeats of the alzheimer-related protein tau. *Nat. Commun.* **2017**, *8*, 275. [CrossRef]
118. Wegmann, S.; Eftekharzadeh, B.; Tepper, K.; Zoltowska, K.M.; Bennett, R.E.; Dujardin, S.; Laskowski, P.R.; MacKenzie, D.; Kamath, T.; Commins, C.; et al. Tau protein liquid-liquid phase separation can initiate tau aggregation. *EMBO J.* **2018**, *37*. [CrossRef]
119. Chen, C.; Ding, X.; Akram, N.; Xue, S.; Luo, S.Z. Fused in sarcoma: Properties, self-assembly and correlation with neurodegenerative diseases. *Molecules* **2019**, *24*. [CrossRef]
120. Kostylev, M.A.; Tuttle, M.D.; Lee, S.; Klein, L.E.; Takahashi, H.; Cox, T.O.; Gunther, E.C.; Zilm, K.W.; Strittmatter, S.M. Liquid and hydrogel phases of prp(c) linked to conformation shifts and triggered by alzheimer's amyloid-beta oligomers. *Mol. Cell* **2018**, *72*, 426–443. [CrossRef]
121. Milles, S.; Salvi, N.; Blackledge, M.; Jensen, M.R. Characterization of intrinsically disordered proteins and their dynamic complexes: From in vitro to cell-like environments. *Prog. Nucl. Magn. Reson. Spectrosc.* **2018**, *109*, 79–100. [CrossRef] [PubMed]
122. Kotaja, N.; Bhattacharyya, S.N.; Jaskiewicz, L.; Kimmins, S.; Parvinen, M.; Filipowicz, W.; Sassone-Corsi, P. The chromatoid body of male germ cells: Similarity with processing bodies and presence of dicer and microrna pathway components. *Proc. Natl. Acad. Sci.* **2006**, *103*, 2647–2652. [CrossRef] [PubMed]
123. Ederle, H.; Dormann, D. Tdp-43 and fus en route from the nucleus to the cytoplasm. *FEBS Lett.* **2017**, *591*, 1489–1507. [CrossRef] [PubMed]
124. Gao, Y.; Tan, L.; Yu, J.T.; Tan, L. Tau in alzheimer's disease: Mechanisms and therapeutic strategies. *Curr. Alzheimer Res.* **2018**, *15*, 283–300. [CrossRef] [PubMed]
125. Mandelkow, E.M.; Mandelkow, E. Tau in alzheimer's disease. *Trends Cell Biol.* **1998**, *8*, 425–427. [CrossRef]
126. Lebouvier, T.; Pasquier, F.; Buee, L. Update on tauopathies. *Curr. Opin. Neurol.* **2017**, *30*, 589–598. [CrossRef] [PubMed]
127. Tiwary, A.K.; Zheng, Y. Protein phase separation in mitosis. *Curr. Opin. Cell Biol.* **2019**, *60*, 92–98. [CrossRef] [PubMed]
128. Jiang, H.; Wang, S.; Huang, Y.; He, X.; Cui, H.; Zhu, X.; Zheng, Y. Phase transition of spindle-associated protein regulate spindle apparatus assembly. *Cell* **2015**, *163*, 108–122. [CrossRef] [PubMed]
129. Fuxreiter, M. Fuzziness: Linking regulation to protein dynamics. *Mol. Biosyst.* **2012**, *8*, 168–177. [CrossRef] [PubMed]
130. Tompa, P.; Fuxreiter, M. Fuzzy complexes: Polymorphism and structural disorder in protein-protein interactions. *Trends Biochem. Sci.* **2008**, *33*, 2–8. [CrossRef] [PubMed]
131. Sharma, R.; Raduly, Z.; Miskei, M.; Fuxreiter, M. Fuzzy complexes: Specific binding without complete folding. *FEBS Lett.* **2015**, *589*, 2533–2542. [CrossRef] [PubMed]

Sample Availability: Not available.

molecules

Article

Sequence-Specific DNA Binding by Noncovalent Peptide–Azocyclodextrin Dimer Complex as a Suitable Model for Conformational Fuzziness

Zulma B. Quirolo [1,2,3,†], M. Alejandra Sequeira [1], José C. Martins [2] and Verónica I. Dodero [1,3,*]

1 Depto. Química—Universidad Nacional del Sur, Bahía Blanca 8000 Argentina
2 NMR and Structure Analysis, Department of Organic and Macromolecular Chemistry, Ghent University, 9000 Ghent, Belgium
3 Faculty of Chemistry, Bielefeld University, Universitätsstr. 25, 33615 Bielefeld, Germany
* Correspondence: veronica.dodero@uni-bielefeld.de; Tel.: +49-(0)521-106-2046; Fax: +49-(0)521-106-156963
† Current affiliation: IFISUR—Depto. Química—Universidad Nacional del Sur, Bahía Blanca 8000, Argentina

Academic Editors: Pier Luigi Gentili and Carmelo Corsaro
Received: 17 May 2019; Accepted: 5 July 2019; Published: 9 July 2019

Abstract: Transcription factors are proteins lying at the endpoint of signaling pathways that control the complex process of DNA transcription. Typically, they are structurally disordered in the inactive state, but in response to an external stimulus, like a suitable ligand, they change their conformation, thereby activating DNA transcription in a spatiotemporal fashion. The observed disorder or fuzziness is functionally beneficial because it can add adaptability, versatility, and reversibility to the interaction. In this context, mimetics of the basic region of the GCN4 transcription factor (Tf) and their interaction with dsDNA sequences would be suitable models to explore the concept of conformational fuzziness experimentally. Herein, we present the first example of a system that mimics the DNA sequence-specific recognition by the GCN4 Tf through the formation of a non-covalent tetra-component complex: peptide–azoβ-CyD(dimer)–peptide–DNA. The non-covalent complex is constructed on the one hand by a 30 amino acid peptide corresponding to the basic region of GCN4 and functionalized with an adamantane moiety, and on the other hand an allosteric receptor, the azoCyDdimer, that has an azobenzene linker connecting two β-cyclodextrin units. The azoCyDdimer responds to light stimulus, existing as two photo-states: the first thermodynamically stable with an E:Z isomer ratio of 95:5 and the second obtained after irradiation with ultraviolet light, resulting in a photostationary state with a 60:40 E:Z ratio. Through electrophoretic shift assays and circular dichroism spectroscopy, we demonstrate that the E isomer is responsible for dimerization and recognition. The formation of the non-covalent tetra component complex occurs in the presence of the GCN4 cognate dsDNA sequence ('5-..ATGA cg TCAT..-3') but not with ('5-..ATGA c TCAT..-3') that differs in only one spacing nucleotide. Thus, we demonstrated that the tetra-component complex is formed in a specific manner that depends on the geometry of the ligand, the peptide length, and the ds DNA sequence. We hypothesized that the mechanism of interaction is sequential, and it can be described by the polymorphism model of static fuzziness. We argue that chemically modified peptides of the GCN4 Tf are suitable minimalist experimental models to investigate conformational fuzziness in protein–DNA interactions.

Keywords: GCN4 mimetic; peptides–DNA; E:Z photoisomerization; conformational fuzziness

1. Introduction

1.1. Conformational Fuzziness and Fuzzy Complexes

Dynamics is inherently related to protein functionality and an essential process in the sophisticated transcriptional machinery. This property generates great flexibility in developing new regulatory circuits where many combinations of activators can regulate a wide variety of genes with different coactivator requirements [1,2]. Dynamics results in conformational ensembles, and indeed many proteins exist simultaneously in different yet functionally relevant conformations. The number and type of the different conformational complexes depend on the cellular environment, such as substrate gradient concentration and the different stress signals. In this sense, this conformational 'fuzziness' is often functionally beneficial because it can add adaptability, versatility, and reversibility to small molecule interactions, protein binding, and thereby ease of regulation not only in protein–protein interactions [2,3] but also in protein–DNA interactions. The description of biological complex systems from their ultimate constituents, i.e., atoms or molecules, is beyond our reach since they have variable patterns, whose recognition is made difficult because of their multiple features, variability, and extreme sensitivity depending on the context [3]. One chemical strategy to tackle this complexity is to obtain tailored biomolecules whose conformation can be controlled externally. Therefore, the conformational fuzziness of the biomolecular interaction can be investigated [4].

Finally, Tompa and Fuxreiter [1] introduced the concept of fuzzy complexes to describe binding situations in which at least one of the elements in the complex remains dynamic. Structural disorder in fuzzy complexes represents a continuum, from rather rigid polymorphic complexes displaying static disorder with only a few alternative conformations to highly dynamic random complexes [1,2,5].

Since up to now, the fundamental understanding of the spatiotemporal control of the interaction between transcription factors (Tfs) and DNA remains elusive, we envisaged designing a suitable Tf biomimetic system combining molecular recognition and supramolecular chemistry to exert the external control of the interaction. We hypothesized that a non-covalent interaction between the Tf and the DNA would be an optimal system to mimic the biological interaction. Under this circumstance, different possible complexes and their respective conformers would exist in equilibrium in the unbound and bound states. Therefore, we anticipate that the concepts of conformational fuzziness and fuzzy complexes would facilitate the analysis of our experimental model. We think that such a system and its analysis would offer a unique strategy to understand the molecular triggers of transcription and the underlying fuzzy mechanism of the Tf–DNA interaction.

1.2. Design of the Non-Covalent Externally Controlled Biomimetic System

In *Saccharomyces cerevisiae*, GCN4 transcription factor (Tf) is part of the "general control" system of amino acid biosynthesis, a network of at least 12 different biosynthetic pathways [6]. GCN4 is a member of the bZIP (basic leucine zipper) family of transcriptional activators that bind to the major groove of double-stranded DNA as a homodimer (Figure 1A) [7,8]. The natural dsDNA specific target sequences of the GCN4 dimers are the activator protein 1 binding site (AP1, 5′-..ATGAcTCAT..-3′) and the related cyclic AMP response element (CRE, 5′-..ATGAcgTCAT..-3′). GCN4 is composed of a C-terminal leucine zipper sequence that associates into non-covalent, parallel, alpha-helical dimers, and an N-terminal basic region necessary for binding DNA. The basic regions are disordered in the absence of DNA and form alpha helices only when as a homodimer binds to the cognate DNA (Figure 1A). Therefore, GCN4 Tf is a suitable protein model to investigate the relationship between structural order/disorder by binding with its mediators or cofactors [9] and in the context of protein–DNA interactions.

Pioneering work in the construction of GCN4 mimetics was performed by Kim et al. [10]. They showed that the replacement of the leucine zipper segment of GCN4 by a cysteine capable of dimerization through disulfide bond formation (GCN4-br1, Figure 1A), allowed specific recognition of its consensus DNA sequences, CRE ('5-..ATGA cg TCAT..-3′) and AP1 ('5-..ATGA c TCAT..-3′), with a nanomolar affinity at 4 °C (Figure 1B). Importantly, the monomeric 34-amino acid sequence was not capable of binding to dsDNA by itself, pointing to dimerization as an essential prerequisite. Furthermore, it was possible to trim the basic region to only 23 amino acids, while maintaining the specific recognition of consensus sequences by the covalent cysteine dimer at affinities around 10 nM at 4 °C [11]. In an alternative approach, Morii et al. [12] used a 23-residue peptide to form a non-covalent heterodimeric complex through host–guest supramolecular interactions. One monomer was equipped with an adamantane group (Ad) and the other with a β-cyclodextrin group (β-CyD). In the presence of the cognate dsDNA, they formed a non-covalent heterodimer. The heterodimer obtained by formation of the inclusion complex between the β-CyD (host) and the Ad (guest) was capable of selectively recognizing the CRE binding domain [12]. To address and control the reversibility of the binding between the GCN4 Tf mimetic and its cognate dsDNA, Mascareñas' group employed an azobenzene moiety covalently attached to the basic region of GCN4 with 26 amino acids [13]. This covalent system was capable of selective binding and recognizing its target dsDNA after UV-light irradiation when the azobenzene was in the (Z) conformation, effectively creating an off–on system. More recently, the same group designed a stimuli-responsive system that targeted different dsDNA triggered by different metals [14]. Building on these previous examples and to address the fuzziness of the Tf–DNA interaction, we envisaged employing an alternative homodimeric non-covalent strategy where the addition of an external molecule would promote not only dimerization but also control over the specific DNA recognition by the geometry of host–guest interactions. In this hypothetical system, the cognate protein–DNA interaction would build a non-covalent tetra-component complex only if the geometry of the host–guest interaction would promote peptide dimerization and a concomitant specific DNA binding occurs. However, many different complexes and their conformers are possible. Considering the non-covalent approach, there are opportunities to control externally the system and to modulate the number and type of complexes and their conformers depending on the geometry of ligand and the dsDNA sequence (Figure 1C).

To address host–guest interactions and switchability of the dimerization, we employed a molecule with two β-cyclodextrin units connected via an azobenzene group, a β-CyD-azobenzene–β-CyD (β-azoCyDdimer). This double host would be capable of binding two basic regions of GCN4 if the peptide were equipped with the adamantane (Ad) guest moiety forming a conjugated peptide–Ad molecule.

In general, double-host β-CyDs bind with higher affinity to Ad than simple β-CyD because of a cooperative effect, as described by Breslow et al. [15]. Moreover, the strategy of a non-covalent β-CyD double host brings the possibility to modulate the geometry and inclusion properties of the peptide–(β-azoCyDdimer) interaction by the photoisomerization of the azobenzene moiety. Thus, the photoswitchable host will work as an allosteric receptor in the petide-DNA interaction. The adamantane moiety was incorporated into two minimal mimetics of GCN4 containing 26 amino acid and a second one with 30 amino acids, SH26 and SH30, respectively (Figure 1A). Both sequences contain a flexible GG linker, and the longer SH30 sequence contains the sequence RMKQ in the C-terminus that should enhance the binding capacity to the cognate dsDNA sequences [9]. To investigate the recognition capacity and the sequence specificity of the peptide–azoβ-CyD(dimer)–peptide–DNA interaction, electrophoretic mobility shift assays (EMSA), and circular dichroism spectroscopy (CD) in the presence of the cognate and non-cognate sequences were tested (Figure 1B). From EMSA experiments the stoichiometry of the interaction, if any, could be identified; meanwhile, the CD would provide us information about the transition from disorder-to-partial-order or disorder-to-order in the bound state by monitoring the alpha helix content of the peptide.

In summary, we hypothesized the formation of a non-covalent tetra-component complex, (peptide–azoβ-CyD(dimer)–peptide–DNA), only if the required geometry to drive dimerization and recognition is provided by the host–guest interaction, which depends ultimately on light illumination (Figure 1C). In this way, new possibilities may arise to investigate the different conformations that are involved in DNA recognition triggered by an external stimulus and evaluate the concept of fuzziness in protein–DNA interactions.

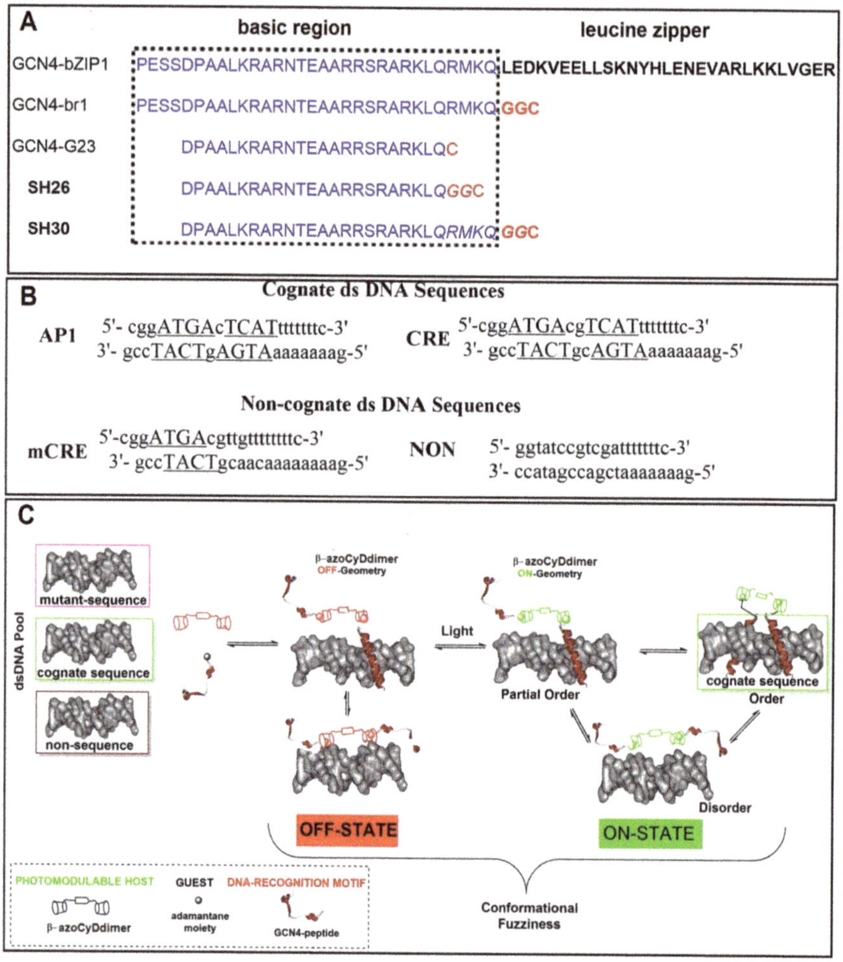

Figure 1. (**A**) Natural GCN4 sequence (GCN4-bzip), and different truncated GCN4 basic regions: GCN4-br1 [8], GCN4-G23 [9], SH26 [11], and SH30. (**B**) Cognate and non-cognate oligonucleotides employed in this work; the recognition and binding sequences are underlined. (**C**) Schematic representation of the possible conformational fuzziness product of the interaction among (β-azoCyDdimer)–peptide–Ad and different dsDNA. By controlling the geometry of the host–guest interaction triggered by light, it would be possible to promote dimerization and the recognition of the cognate dsDNA sequence. The structural transitions of the dimerization and recognition regions upon different partner interactions generate a structural and dynamical continuum towards the formation of the specific non-covalent tetra-component complex *only* in the case of the cognate dsDNA. Although other complexes with different stoichiometry can be formed, they are not shown to simplify the scheme.

2. Results

2.1. Synthesis and Characterization of the Allosteric Receptor and Adamantyl Substituted Peptides

2.1.1. Allosteric Receptor Synthesis and Characterization

Herein, we simplified the synthesis of the β-CyD double host, azoCyDdimer [16] (Figure 2A) by direct coupling of 4,4′-bis (carboxy) azobenzene acid and mono-6-amino-6-deoxy-β-cyclodextrin (β-CyDNH$_2$) in DMF using HATU, as a coupling agent and DIPEA. The reaction was monitored by TLC until the disappearance of β-CyDNH$_2$, and purified by preparative HPLC obtaining 20% yield. In general, monosubstituted cyclodextrins (CyD) lack the high symmetry of free or symmetrically substituted CyD and, therefore, present substantially more complex NMR spectra. Previously, the azoCyDdimer was not entirely characterized [14]. Taking this into account, we present here the complete characterization of the azoCyDdimer by a combination of 2D NMR experiments. The azoCyDdimer structure, made by two CyD linked through an azobenzene unit, was initially studied by ^{1}H NMR in DMSO (Figure 2A,C, and Supplementary Materials, Figure S1). Two distinct anomeric proton signals at δ = 4.96 ppm for H-1 and δ = 4.83 ppm for H-1′ were observed (with a relative area of 12:2), corresponding to the pattern of monosubstituted CyD. The –OH2 and –OH3 groups were observed between 5.70 to 5.82 ppm. The other group of signals between 3.20 and 3.80 ppm corresponded to the rest of the protons of the glucose units and suffered from substantial overlap (see Supplementary Materials, Figure S1). Nevertheless, the complete assignment could be obtained through the use of COSY, TOCSY, ^{1}H-{^{13}C} HSQC, and HMBC.

Regarding the signals of the photomodulable connector, two doublets of 4H area were observed at δ = 7.98 (J = 8.6 Hz) and δ = 8.04 (J = 8.6 Hz), which corresponded to a substituted azobenzene derivative in position 4.4 ′ (isomer (E)). Using a multiplicity edited HSQC experiment, CH units were differentiated from CH$_2$, while the absence of correlations allowed the identification of signals corresponding to the NH and OH groups. The analysis was completed by HMBC, which allowed establishment of longer distance correlations between ^{1}H and ^{13}C over two or three bonds. This enabled the correlation of Hb at 8.00 ppm (doublet) with the carbonyl of the amide, while a weak correlation was also observed between Cb and the H of the amide thus confirming its integrity (see Supplementary Materials, Figure S2 and S3).

Furthermore, we evaluated the E:Z photoisomerization of the dimer using ^{1}H NMR and UV–Vis spectroscopy (Figure 2C and see Supplementary Materials, Figure S4). Changes in the E:Z ratio after photoisomerization were calculated from the integration of the corresponding areas of the azobenzene protons, as shown in Figure 2C. The signals corresponding to the Z isomer appeared at the higher field than the corresponding signals of the E isomer due to the different contribution of the diamagnetic anisotropy of both benzene rings in the E:Z isomers. As mentioned, the ^{1}H-NMR spectrum of the E isomer (Figure 2(C1)) had two doublets at 8.03 ppm (J = 8.6 Hz) and 7.97 ppm (J = 8.6 Hz) corresponding to the aromatic protons Hc and Hb of the azobenzene linker group, and a singlet at 8.52 ppm corresponding to the protons of the amide group. After the isomerization, a decrease of these signals was observed while three new signals corresponding to the Z isomer appeared: two doublets at 7.72 ppm (J = 8.4 Hz) and 6.90 ppm (J = 8.4 Hz) and a singlet at 8.20 of the H corresponding to the amide bond (Figure 2(C2)).

Previously, it was reported that azoCyDdimer forms supramolecular aggregates in water above the concentration 1 mM, while below this concentration it reaches the photostationary state of E:Z 60:40 as determined by UV–Vis spectroscopy [16]. Here, we analyzed the dependence of the concentration and isomerization by the integration of the peaks using ^{1}H-NMR in DMSO. We observed that the most diluted samples favored the formation of the Z isomer, from E:Z ratios of 40:60 (0.5 mM) to 60:40 (15 mM) (Figure 2D,E).

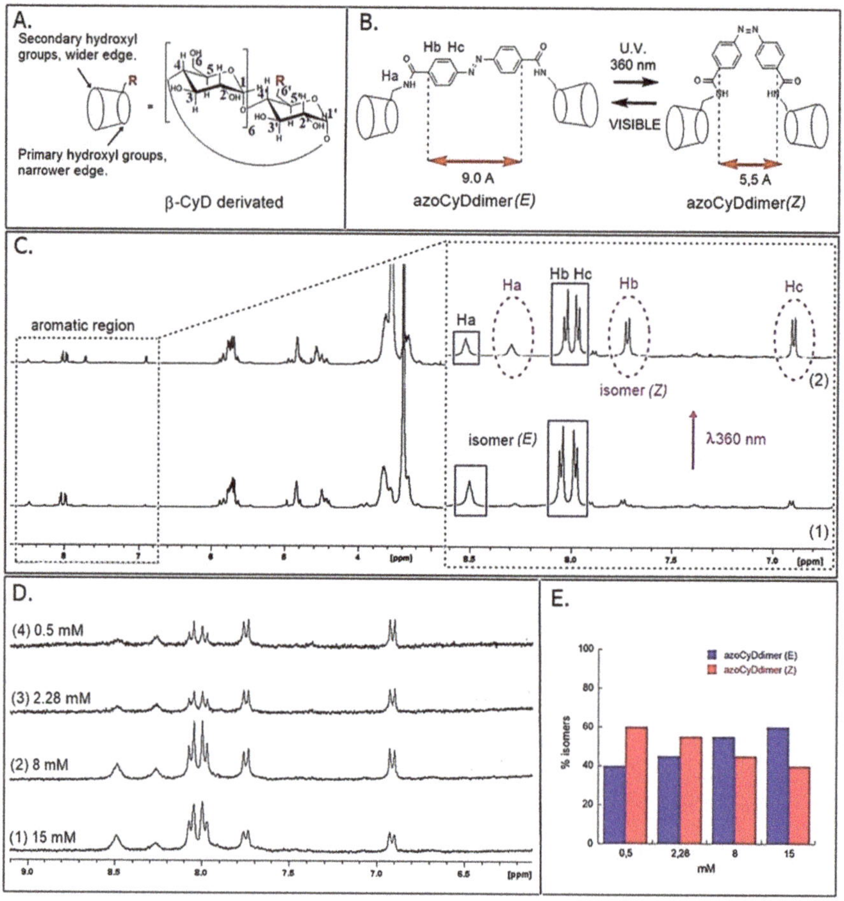

Figure 2. (**A**) Structure of the allosteric receptor. (**B**) Representation of the photoisomerization of the azoCyDdimer. (**C1**) ^{1}H-NMR-500MHz spectra of azoCyDdimer (15 mM) at 25 °C (DMSO-d_6). (**C2**) ^{1}H NMR spectra of azoCyDdimer after 4 h irradiation with UV light (λ = 360 nm). Isomer (E) signals are marked with full lines and isomer (Z) signals with dotted lines. (**D**) Independent ^{1}H-NMR (300MHz) experiments of the photoisomerization of azoCyDdimer at different initial concentrations, showing the aromatic region after 20 min of irradiation at 360 nm in DMSO-d6, 25 °C (see Materials and Methods). (**E**) Bar graph of E:Z isomer ratio obtained by integration of the ^{1}HNMR signals in D.

2.1.2. Peptide Derivatives Synthesis

To obtain a minimal GCN4 mimick, we synthesized two adamantane GNC4 derivatives (Ad26 and Ad30, Figure 3A). Both peptides were obtained by coupling the bromoacetyladamantane with the thiol group of Cys26 or Cys30, respectively.

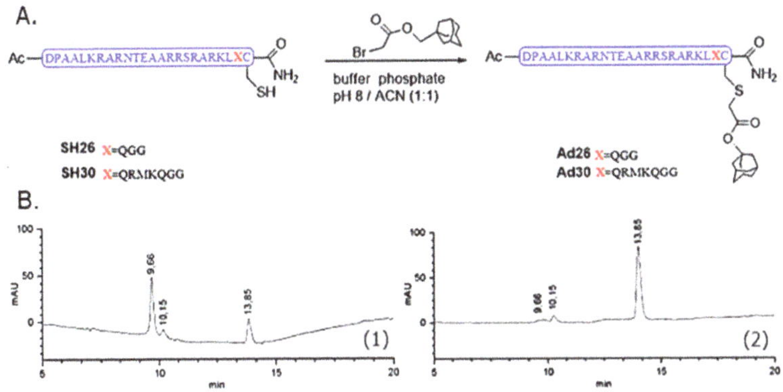

Figure 3. (**A**) GNC4 peptides derivatives. (**B**) RP-LC–MS-monitoring of the reaction between SH30 and bromoacetyl adamantane to obtain Ad30 at 5 (1) and 120 (2) min.

We took advantage of the nucleophilic reactivity offered by the thiol group of cysteine in a solution using the completely deprotected peptide. The synthesis of the new Ad30 was performed following the established protocol for Ad26 [17,18]. In Figure 3B, LC–MS chromatogram shows the evolution of the reaction, which was verified by ESI. After 2 h of reaction, the peak of SH30 (tr 9.56 min) disappeared and a new peak at 13.85 min appeared, whose m/z corresponded to the final product Ad30.

Both Ad26 and Ad30 peptides were employed as the minimal GCN4 derivatives to evaluate the interaction strategy of non-covalent homodimers. Furthermore, we employed the corresponding disulfur dimers SS52 and SS60 as positive controls of the peptide–DNA interaction [10,17,18].

2.2. *Peptide–dsDNA Interaction*

2.2.1. Electrophoretic Mobility Shift Assays (EMSA)

When a GCN4 peptide binds to a cognate dsDNA, it induces a delay in the migration of the peptide–dsDNA complex in comparison with free dsDNA when using EMSA. This interaction leads to a characteristic shift in the position of the associated band, leading to the observation of two bands with an intensity ratio reflecting the equilibrium involved in complex formation. EMSA experiments require a minimal amount of sample in the nM range, making it an ideal and common technique to obtain qualitative information of the respective peptide–DNA interaction [10–13,17–19]. The DNA-binding properties of the synthetic peptides were studied by EMSA under non-denaturing conditions and using SYBR-gold for DNA staining. Initially, the incubation of Ad26 in the presence or absence of azoCyDdimer to both AP1 and CRE dsDNA (Figure 1B) showed no interaction bands, as shown in Figure 4. In this case, a new band could only be observed for the positive control SS52, which is the SH26 disulfide dimer (Figure 4, lane 2), confirming the negative result. It was noticeable that the interaction of Ad26 with CRE (Figure 4C,D) produced a decrease in the free DNA band. We hypothesized that this behavior might be a consequence of the formation of different types of complexes with the dsDNA, which would make diffuse broadband under both conditions. Nevertheless, no strong interaction was detected at the nM range. The azobenzene molecule extends over a considerably different distance when considering the E and Z isomers, being theoretically about 9 to 5.5 Å, respectively [20–22]. When we tried to mimic the binding system of GCN4, this distance may not have been adequate for the specific interaction with the short sequence and its putative dsDNA binding site. Considering that no new migration band was observed in the nM, this system was not suitable for further evaluation, mainly because upon increasing peptide concentration the unspecific interactions started to be relevant.

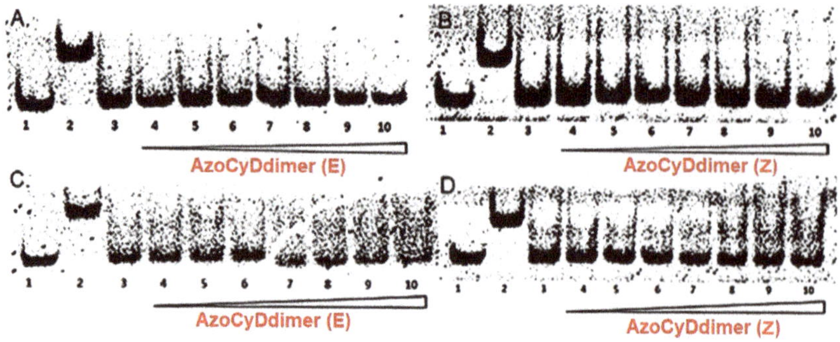

Figure 4. EMSA analysis of DNA binding of Ad26 peptide in the presence and absence of azoCyDdimer. (**A**) Lanes 1–10: 50 nM AP1 ('5-cggATGA c TCATtttttttc-3'); lane 2: 300 nM SS52; lanes 3–10: 200 nM Ad26 and 0, 0.5, 1, 2, 5, 20, 50, 100 equivalents azoCyDdimer (E). (**B**) Lanes 1–10: 50 nM AP1 ('5-cggATGA c TCATtttttttc-3'); lane 2: 300 nM SS52; lanes 3–10: 200 nM Ad26 and 0, 0.5, 1, 2, 5, 20, 50, 100 equivalents azoCyDdimer (Z). (**C**) Lanes 1–10: 50 nM CRE ('5-cggATGA cg TCATttttttt-3'); lane 2: 300 nM SS52; lanes 3–10: 200 nM Ad26 and 0, 0.5, 1, 2, 5, 20, 50, 100 equivalents azoCyDdimer (E). (**D**) Lanes 1–10: 50 nM CRE ('5-cggATGA cg TCATtttttttc-3'); lane 2: 300 nM SS52; lanes 3–10: 200 nM Ad26 and 0, 0.5, 1, 2, 5, 20, 50, 100 equivalents azoCyDdimer (Z). DNA was detected by fluorescent staining with SYBR-gold.

On the other hand, incubation of the derivative Ad30 with AP1 showed a new migration band, the molecular weight of which corresponded to the Ad30 monomer that binds to dsDNA (Figure 5A,B, lane 2, compare with the positive interaction with the covalent dimer of SH30, named as SS60) [12,17]. It seems that the monomer Ad30 interacted with one half of the recognition site, but no tri or tetra-component complex was obtained, leading us to conclude that this interaction should be considered as unspecific, although with high affinity. Interestingly, once the azoCyDdimer was present in the incubation mixture with Ad30 and the CRE sequence, two bands appeared, one corresponding to the above-mentioned unspecific interaction, and the second one with a retardation band compatible with a higher order complex (Figure 5C,D). By increasing the concentration of azoCyDdimer, the formation of the second complex was favored. This new migration band was compatible with the size of the tetra-component system Ad30–AzoCyDdimer–Ad30–CRE that migrated less than the disulfide dimer (compare with lane 2, Figure 5). The intense band of specific binding was only observed in the presence of an excess of azoCyDdimer enriched in the isomer (E), to the detriment of the nonspecific binding that nearly disappeared at the highest azoCyDdimer concentration (Figure 5C, lanes 10–12). A similar result was observed when the azoCyDdimer (Z) was added to the incubation mixture; however, in this case, the specific band was more diffuse, showing the existence of different complexes in the mixture (Figure 5D, lanes 9–12).

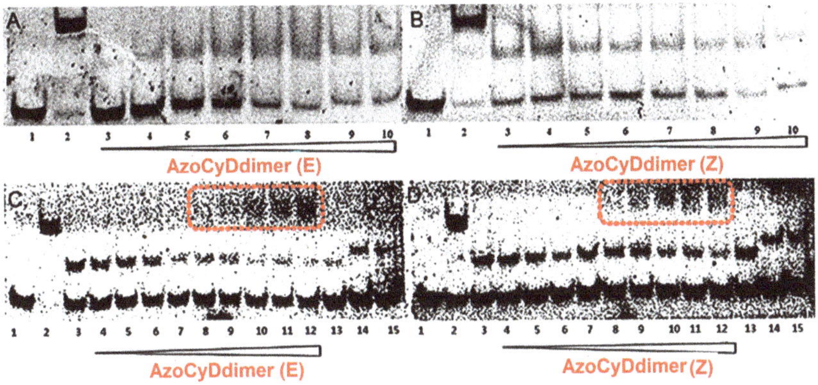

Figure 5. EMSA analysis of DNA binding of Ad30 peptide in the presence and absence of azoCyDdimer. (**A**) Lanes 1–10: 50 nM AP1 ('5-cggATGA c TCATtttttttc-3'); lane 2: 300 nM SS60; lanes 3–10: 200 nM Ad30 and 0, 0.5, 1, 2, 5, 20, 50, 100 equivalents azoCyDdimer (E). (**B**) Lanes 1–10: 50 nM AP1 ('5-cggATGA c TCATtttttttc-3'); lane 2: 300 nM SS60; lanes 3–10: 200 nM Ad30 and 0, 0.5, 1, 2, 5, 20, 50, 100 equivalents azoCyDdimer (Z). (**C**) Ad30 and azoCyDdimer (E): lane 1–13: 50 nM CRE ('5-cggATGA cg TCATtttttttc-3'); lane 2: 100 nM SS60; lanes 3–15: 200 nM Ad30; lanes 3–10: azoCyDdimer (E): 0, 0.5, 1, 2, 10, 50, 100, 500 eq.; lane 11–12: excess of azoCyDdimer (E) (0.15 mM); lane 13: Ad30 200 nM; lane 14: 50 nM ds mCRE ('5-cggATGAcgttgtttttttc-3'): 200 nM Ad30; lane 15: 50 nM mCRE ('5-cggATGAcgttgtttttttc-3'): 200 nM Ad30, 0,5 eq azoCyDdimer(E). (**D**) Ad30 and azoCyDdimer (Z) lane 1–13: 50 nM CRE ('5-cggATGA cg TCATtttttttc-3'); lane 2: 100 nM SS60; lane 3–10: 200 nM Ad30 and azoCyDdimer (Z): 0, 0.5, 1, 2, 10, 50, 100, 500 eq.; lane 11–12: Ad30: 200 nM and excess of azoCyDdimer (Z) (0.15 mM); lane 13: Ad30 200 nM; lane 14: 50 nM ds mCRE ('5-cggATGAcgttgtttttttc-3'); 200 nM Ad30; lane 15: 50 nM ds mCRE ('5-cggATGAcgttgtttttttc-3'): 200 nM Ad30, 0,5 eq. azoCyDdimer (Z). DNA was detected by fluorescent staining with SYBR-gold.

Additionally, in the presence of the azoCyDdimer (Z), the relative intensity of the non-specific band did not show a significant decrease in comparison with the specific one. To further investigate the specificity of the interaction, we employed a dsDNA with half CRE sequence (mCRE), (Figure 5, in lanes 14 and 15, gels C and D) showing the migration of the band compatible with a complex monomer Ad30–azoCyDdimer–mCRE. The slightly different migration profile could be a different migration of both dsDNA sequences. It is important to mention that under this circumstance, it seems that the mixture enriched in E isomer (95:5) could form the cooperative complex. Therefore, the mixture enriched in the Z isomer, which still contained 60% of E isomer after photo-illumination, could also bind.

Considering the positive binding of the Ad30–azoCyDdimer–Ad30 system and the CRE sequence, we further evaluated its interaction by circular dichroism (CD) spectroscopy.

2.2.2. Circular Dichroism Spectroscopy

There is a proportional relationship between the amount of α-helix and the intensity of the negative signal at 222 nm in the circular dichroism spectrum, especially useful for study the bZIP-DNA interactions [23,24]. In the case of GCN4, which is disordered in the absence of its cognate dsDNA, there is an increase in its folding from random coil to α-helix when it interacts specifically with its consensus DNA. Briefly, Ad30 solution was added to a 5 μM solution of azoCyDdimer (E) in phosphate buffer to monitor any change due to the interaction. A decrease of the band at 222 nm from −0.2118 (CyDdimer only) to −16.201 °cm^{2}dmol^{-1} was observed when Ad30 was added (Figure 6A). In the presence of the CRE sequence ('5-..ATGA cg TCAT..-3'), a decrease of the same band was observed because of the interaction with the CRE binding site at −30,666 °cm^{2}dmol^{-1}, validating the specific interaction of the Ad30–azoCyDdimer (E)–Ad30 system and CRE obtained by EMSA assays (Figure 5C).

The sample was irradiated with UV light (360 nm) to increase the ratio of isomer Z in the photostationary state, and a slight change in the band at 222 nm was observed, from −29.968 °cm^{2}dmol^{-1} at 20 min to −28.953 °cm^{2}dmol^{-1} at 40 min. As in the case of isomer (E), the solution enriched with azoCyDdimer (Z) did not show a significant contribution of the ellipticity at 222 nm (−0.2956 °cm^{2}dmol^{-1}) (Figure 6B). With the addition of the solution of Ad30 peptide, the value of the band at 222 nm decreased to −15.797 °cm^{2}dmol^{-1} and then when CRE was added, the second decrease in ellipticity was observed, reaching the value of −26.709 °cm^{2}dmol^{-1}, which showed lower alpha-helical content in comparison with the azoCyDdimer (E). Once this mixture was irradiated with white light, the signal at 222 nm decreased slightly to −27.034 °cm^{2}dmol^{-1}.

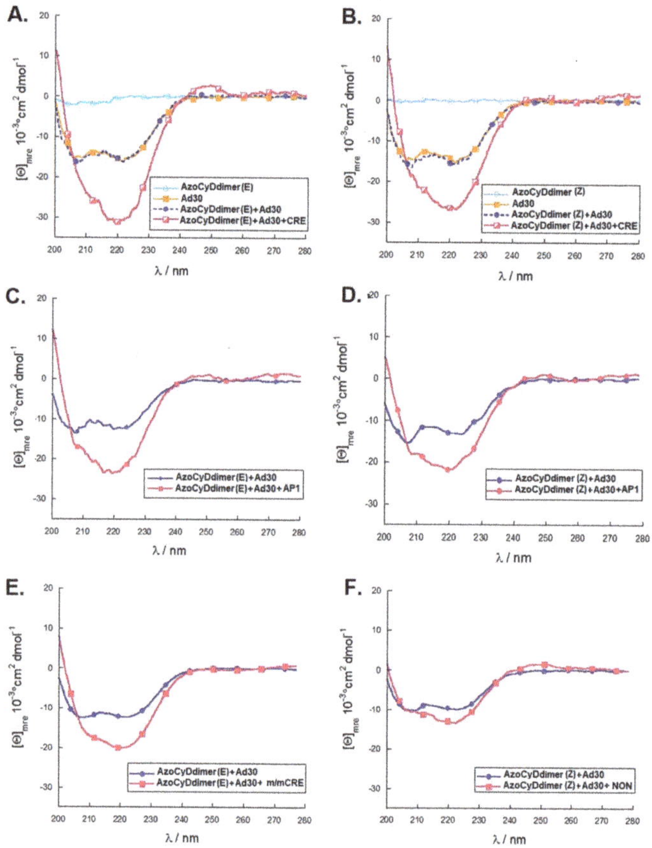

Figure 6. Circular dichroism evaluation of the interaction of Ad30 (**A**) with azoCyDdimer (E) and its cognate sequence CRE ('5-cggATGA cg TCATtttttttc-3'), (**B**) with azoCyDdimer (Z) and its cognate sequence CRE ('5-cggATGA cg TCATtttttttc-3'), (**C**) with azoCyDdimer (E) and its cognate sequence AP1 ('5-cggATGA c TCATtttttttc-3'), (**D**) with azoCyDdimer (Z) and its cognate sequence AP1 ('5-cggATGA c TCATtttttttc-3'), (**E**) with azoCyDdimer (E) and mCRE ('5-cggATGAcgttgtttttttc-3'), and (**F**) with azoCyDdimer (E) in the presence of a random sequence, NON ('5-ggtatgcgtcgattttttttc-3'). In all experiments, the band of dsDNA was subtracted; for further experimental details refer to Materials and Methods section.

On the other hand, in the interaction of both isomers with AP1, ('5-..ATGA c TCAT..-3'), the band at 222 nm decreased to −22.974 °cm^{2}dmol^{-1} in the case of the isomer E and to −21.408 °cm^{2}dmol^{-1}

when the isomer Z was present (Figure 6C,D). Both results were in the same direction as the EMSA, showing that the geometry of the dimer was not favorable for the interaction with AP1, which had only one base spacer between both recognition sites.

Finally, we tested the unspecific interaction with the half sequence of mCRE ('5-..ATGA cg..-3'), detecting a decrease of the molar ellipticity to −20.214 °cm^{2}dmol^{-1}. This value, as expected, was lower than the one observed with the full sequence (compare Figure 6A,E). The observed decrease in molar ellipticity could be due to the insertion of the Ad30 monomer in the major groove and interaction with the bases of the half sequence ('5-..ATGA cg..-3'). To investigate the role of the phosphates of the DNA and their binding to the basic Ad30, we experimented with a random DNA sequence that does not possess a cognate binding site, named NON ('5-ggtatgcgtcgattttttc -3') (Figure 6F). For this case, a slight decrease in the signal from −10.389 to −14.323 °cm^{2}dmol^{-1} was observed. From that experiment, we can exclude that the unspecific interaction with the negative phosphates of DNA and the basic GCN4 peptide has a role in the interaction of Ad30.

These results are in agreement with those obtained by EMSA, showing that Ad30 in the presence of azoCyDdimer in the E configuration presented a specific binding to CRE sequence forming the tetra-component complex; meanwhile, in the Z configuration, the helical content was lower. Considering that the E isomer is an equilibrated mixture 95:5 (E:Z) and the Z isomer is a 60:40 (E:Z) mixture, we hypothesized that the E isomer has the right geometry to form the specific tetra-component complex. The CD signal of the interaction with AP1 was in the same range of those observed for the half-recognition sequence, mCRE. This shows the feasibility of a monomer Ad30–AzoCyDdimer to bind to the sequence ('5-..ATGAcg..-3'), but not as a dimer as detected by the EMSA experiment (Figure 5A,B). Importantly both CD and EMSA experiments demonstrated the feasibility of formation of the non-covalent complex Ad30–AzoCyDdimer (E)–Ad30 that interacts preferably with CRE sequence, ('5-..ATGA cg TCAT..-3').

2.2.3. Competitive Binding Assay with 1-Adamantane Acetic Acid by CD and EMSA Experiments

To further corroborate the formation of the inclusion complex Ad–βCyD in the system Ad30–AzoCyDdimer (E)–Ad30, we hypothesized that such interaction would collapse upon addition of an excess amount of a competitive β-cyclodextrin binder, such as 1-adamantane acetic acid [12]. Under this circumstance, the helical content would decrease due to the dimer collapse, and the specific band in the EMSA experiment should disappear. Following this hypothesis, we evaluated first the system by the EMSA experiment (Figure 7A).

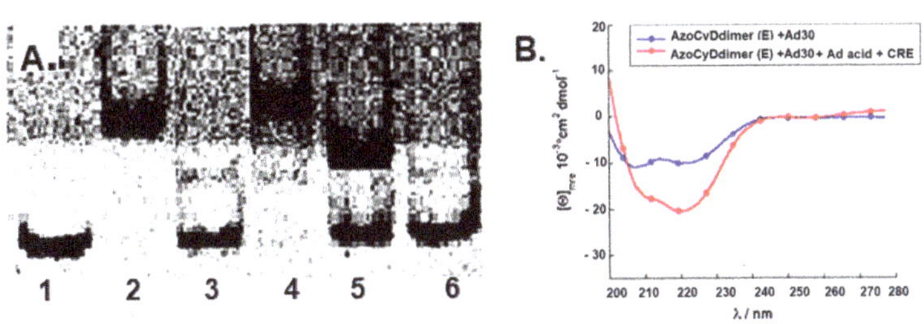

Figure 7. (**A**) EMSA analysis of DNA binding of Ad30–azoCyDdimer (E)–Ad30 in the absence and presence of 1-adamantane acetic acid. Lanes 1–5: 50 nM CRE ('5-cggATGA cg TCATttttttc-3'); lane 2: 300 nM SS60; lane 3: 200 nM Ad30; lane 4: 200 nM Ad30 and 100 equiv azoCyDdimer (E); lane 5: 200 nM Ad30, 100 equiv azoCyDdimer (E) and 200 equiv 1-adamantane acetic acid; lane 6: 50 nM NON, 200 nM Ad30, 100 equiv of azoCyDdimer (E) and 200 equiv 1-adamantane acetic acid. (**B**) CD spectra of the interaction Ad30–AzoCyDdimer (E)–Ad30–CRE in presence and absence of 1-adamantane acetic acid.

It was observed that the tetra-component complex (lane 4) in the presence of an excess of the 1-adamantane acetic acid was not formed (lane 5); instead, only the unspecific shift band of the complex peptide:DNA (1:1) was detected. In the case of the interaction of the NON, any shift band was detected in the presence of the competitive β-cyclodextrin binder (lane 6). By CD, it was observed that in the presence of an excess of 1-adamantane acetic acid, the negative band at 222 nm reached only the value of −20.409 °cm^{2}dmol^{-1} instead of the expected −30,666 °cm^{2}dmol^{-1} for the dimer (Figure 7B). The obtained value was in the range of the unspecific interaction obtained by interaction with the half-sequence mCRE sequence (−20.214 °cm^{2}dmol^{-1}) (see Figure 6E). These results confirm the relevance of the βCyD–Ad host–guest interaction to promote dimerization and thus trigger the specific interaction with the target dsDNA.

3. Discussion

Transcription factors (Tfs) are useful proteins to evaluate the concept of fuzziness in protein–protein and protein–DNA interactions [1,2,7,25–29]. For GCN4 Tf, it has been hypothesized that the binding to DNA occurs sequentially in such a way that a monomer is first assembled into the major groove of DNA, followed by dimerization with a second GCN4 Tf unit [30]. Generally, mimetics of the GCN4 Tf possess a high degree of randomness, but in the presence of a dimerization motif and their cognate ds DNA, there is a substantial increase of α helical structure due to structure inducing complex formation as a result of the specific binding [9–14]. Conformational dynamism and heterogeneity enable context-specific functions to emerge in response to changing environmental conditions and, furthermore, allow a single structural motif to be used in multiple settings [14]. The sequential interaction pathway is a natural strategy that prevents dimers from being trapped for a relatively long time in non-consensus sequences. In the case of GCN4 mimetics, the dimerization motif (Figure 1A), which is not involved in the DNA binding interface, is generally conformationally unaffected by binding to the DNA. Thereby, it retains the capability to modulate the interaction [31]. In general, heterogeneous conformational segments can increase binding affinity. This conformational flexibility and heterogeneity of proteins represent their fuzziness [1–6]. The classical framework of protein interactions establishes that there is a deterministic relationship between protein sequence and function. Based on this, a distinguished three-dimensional arrangement of the amino acids is a prerequisite for a given biological activity and is unambiguously encoded in the sequence [32]. However, protein functions are modulated by different mechanism triggered by different effectors. The effector perturbs one site and thereby leads to altered activity in a second, substrate site [33]. In our system, the effector was the azoCyDdimer, which triggered host–guest interaction of β-CyD of the dimer and two GCN4 peptides containing each an adamantane moiety. The different geometry the azoCyDdimer was previously calculated as 9.1 Å and 6.6 Å for E and Z isomers, respectively [16]. As confirmed here, the azoCyDdimer exists under two photo-states: the first thermodynamically stable with an E:Z isomer ratio of 95:5 and the second obtained after irradiating with ultraviolet light, E:Z 60:40 ratio. Taking into account, the large geometrical difference between both isomers and the relevance of this distance to form a suitable complex, we had hypothesized that it would be unlikely that both isomers could form a homodimer leading to the same specific dsDNA interaction. Considering that the first and second photo-stationary state contains a major proportion of E isomer 95% and 60%, respectively, it can be anticipated that if the E isomer is the responsible of the tetra-component complex, the specific interaction would be observed under both illumination conditions. However, some differences in binding affinity can be expected. On the contrary, if the Z isomer would give the right geometry for the peptide–DNA interaction, the positive interaction would be observed only mainly in the second photostationary stage where the Z isomer is present in a significant concentration (40%). Nevertheless, if both isomers would contribute equally to the tetra-component complex, similar behavior in all the experiments would be expected.

We performed two complementary sets of experiments, EMSA and CD, to evaluate the system. As aforementioned, from EMSA experiments the stoichiometry of the interaction, if any, was identified. Meanwhile, the CD gave information about the transition from disorder-to-order and disorder-to-partial-order in the bound state. In such a case, the value of the molar helicity per residue would let us set a scale of the order obtained for each interaction and connecting them with the stoichiometry obtained in the EMSA experiment. Considering the theory of fuzzy complexes [1], the disordered-binding elements of GCN4 mimetic may undergo three types of structural transitions upon interaction with the target dsDNA. First, let us consider the disorder-to-order transition to adopt a stable, well-defined conformation in the bound state [32]. This is also referred to as coupled folding to binding [34]. Second, upon partner recognition, a transition from disorder-to-partial order could take place in shallow, often hydrophobic binding pockets. In this case, the interface was generated by many redundant contacts and few specificities, as observed by the formation of the complex DNA–Ad30 and DNA–Ad30–azoCyDdimer (E) [32]. Third, a disorder-to-disorder transition may occur upon binding of the Ad30–azoCyDdimer–Ad30 to the cognate dsDNA; however, in such case, we would not expect an increase of the helical content, but binding should be observed in the EMSA experiment.

We found that Ad30, but not its shorter form Ad26, was able to form two detectable new interaction shift bands during EMSA experiments. The shifted band detected in the absence of azoCyDdimer has a migration compatible with the monomer of Ad30, which binds to all the dsDNA sequences, inclusive to the random one (NON). Thus, this interaction was considered as unspecific. The second shift band was observed in excess of azoCyDdimer, and it corresponds to the tetra-component complex. This second band of the non-covalent tetra-component complex was observed only in the presence of CRE ('5-..ATGA cg TCAT..-3') sequence under both illumination conditions, but not with AP1 ('5-..ATGA c TCAT..-3'), nor with mCRE ('5-..ATGAcg..-3'). The elongation of the sequence in Ad30 by insertion of the affinity enhancing QRMK sequence in before the C-terminal QGGC end, thus leading to the QRMKQGGC sequence (Ad30) instead of QGGC (Ad26), does increase the binding affinity to levels allowing the detection of the interaction of one Ad30 monomer with dsDNA containing the sequences ('5-..ATGAc..-3'). Probably the positive charge of the arginine (R) and lysine (K) groups are stabilizing effects that favor the observed interaction. In general, dynamics of the interaction may vary in a wide range depending on the truncation of the disordered dimerization domain as observed here [32]. Considering our findings, we hypothesized that the formation of a monomeric complex Ad30–azoCyDdimer and the ('5-..ATGAc..-3') sequence might favor the subsequent tetra-component complex formation. This can be explained by the fact that when one of the sites is bound to its cognate site receptor, a second site located close-by binds cooperatively, basically because of the lower entropic cost of a (pseudo)-intramolecular interaction [35]. Considering this concept of multivalency, if the linker connecting the binding and the dimerization domains are flexible, the average distance between the sites is the main factor determining the cooperativity. In such complex, we hypothesized that the QRMKQGGC in Ad30 sequence provides the required flexibility to favor the observed tetra-covalent binding and that this region is responsible for the fuzzy behavior of the mimetic in the bound state.

Considering that the specific migration band with CRE was also observed in both photostationary states where the E isomer is 95% and 60%, we hypothesized that the geometry of azoCyDdimer (E) is the adequate combination causing a migration band compatible with the formation of a stable tetra-component complex in the presence of ds CRE. When we evaluated the interaction of Ad30, the azoCyDdimer (E), and the CRE sequence using CD, we observed a significant increase of the helical content compatible with the formation of a tetra-component system, Ad30–azoCyDdimer (E)–Ad30–CRE, similar to those reported by the covalent dimer SS60 and other GCN4 mimetics [10–14,17,18]. In the case of the second photostationary state (ps) where the ps is E:Z (60:40), a moderate and lower increase of Ad30 helical content compared with the first pss E:Z (95:5) was observed, suggesting that the Z isomer does not contribute to the specific binding.

Importantly, for the interaction of Ad30–azoCyDdimer (E) and the mCRE sequence ('5-..ATGAcg..-3'), which contains only one binding site instead of the two necessary ('5-..ATGA cg TCAT..-3') for the tetra-component formation, a much lower increase of the helical content was observed (Figure 6E). This is compatible with the interaction of one Ad30–azoCyDdimer (E) or only Ad30 with this sequence, as detected in the EMSA experiment (Figure 5C,D, lane 15). The interaction with the random sequence (NON, Figure 6F), which has no recognition motif, showed nearly no interaction, demonstrating that the interaction of the basic region with the phosphate backbone of the dsDNA has at most a minor contribution to the protein–DNA binding. Summarizing the CD experiments, the increase of the helical content upon binding is related to the bound state of Ad30–azoCyDdimer (E) with the different dsDNA as follow: CRE > AP1~ mCRE > NON. After comparison of the increase of helical content (order) of Ad30- in the presence of CRE and the two azoCyDdimer isomers, it is evident that the photo-state (E) contribute more than the (Z). It seems that dimerization interaction between the adamantane and β-cavity of the azoCyDimer occurs at the wider edge of the cyclodextrin (Figure 2A), instead of its narrow edge, which would increase the binding affinity [15]. Previously, it has been shown by molecular dynamic simulations that while the Z azoCyDdimer isomer makes a 1:1 complex with a small organic molecule with two adamantane moieties, through interaction with the narrow edge; the E azoCyDdimer isomer forms supramolecular head to tail complexes through the wider edge [16]. In the case of the mimetic of GCN4, such supramolecular aggregates are not possible because the peptide has only one binding domain. All our experiments suggested that E conformation provides the required geometry to favor the tetra-component complex, which depends on the dsDNA sequence. On the other hand, we could not discard the formation of the Ad30–azoCyDdimer (Z)–Ad30. Nevertheless, our findings indicate that the geometry of the interaction in the Z conformation is not adequate for DNA recognition.

Finally, to confirm that the host–guest interaction between the β-CyD and the Ad moiety is a necessary requisite to recognize CRE sequence specifically, we performed a competitive assay in the presence of 1-adamantane acetic acid. Upon addition of the competitive β-CyD, the only band observed was the unspecific 1:1. A comparable result was obtained by CD upon addition of the competitive β-CyD reaching a helical content value similar to the one of the unspecific binding to mCRE. These final experiments sustain the hypothesis that the azoCyDdimer is necessary to form the homodimeric complex leading to the formation of the tetra-component system in a sequence-specific manner. We had planned to modulate the peptide–DNA interaction by the structural change promoted by the photoisomerization of the azobenzene group located as a linker between the two β-CyDs moieties in the azoCyDdimer. However, as well as for the covalent version of photomodulable GCN4 mimetic reported by Caamaño et al., it was not possible to switch the interaction effectively by changing the azobenzene conformation in situ [13]. We could not anticipate that the E isomer would be the one that favors the dimerization. Nevertheless, we have the off situation in the absence of azoCyDdimer switching the recognition to *on* above a certain azoCyDdimer (E) concentration (Figure 8). About the mechanism of interaction, we hypothesized that recognition occurs through a sequential mechanism where the monomer Ad30 probably forming a 1:1 complex with the azoCyDdimer recognizes the sequence ('5-..ATGAc..-3') Initially with high affinity followed by the cooperative union of a second monomer promoted by the host–guest dimerization motif of azoCyDdimer in the E conformation. This second stage is stabilized by the specific interaction with CRE sequences ('5-..ATGA cg TCAT..-3'). During this interaction, different fuzzy conformers are possible that favor the specific interaction with CRE by a transition from disorder-to-partial order of the four components of the complex (Figure 8). For the others dsDNA that lack of the complete binding sequence in mCRE or only one base as a connector between the binding site in AP1, let to the formation of bi-or tri-component complexes. As described before, Tompa et al. have proposed a categorization for fuzzy complexes, as static or dynamic [1]. Considering that in the bound state, most of the Ad30 is order, it is possible to describe the interaction by the polymorphism model of static fuzziness [1]. In such model, one part of the molecule makes the contacts for the interaction whereas its dimerization domain adopts several distinct conformations, and establish the right geometry to favor dimerization and binding to the DNA with the complete recognition sequence in CRE. We hypothesized that the QRMKQGGC region of Ad30 might adopt multiple unrelated conformations to stabilize the interaction with the larger azoCyDdimer (E) instead of the compact (Z) isomer, thus favoring dimerization that contributes to CRE recognition. Probably, this structural variability limits an unfavorable decrease in entropy accompanying complex formation, which enables the combination of rapid and thermodynamically favorable binding. Truncation of this region led to no-binding as observed for Ad26, which validates our hypothesis. In conclusion, the present report is the first example of a GCN4 mimetic that forms a specific non-covalent tetra-component system with the cognate binding sequence only in the presence of an external ligand in one of two possible conformations, working as an off–on switch. Moreover, we demonstrated that chemically modified mimetics of GCN4 are suitable minimalist models to investigate conformational fuzziness in protein–DNA interactions, opening the opportunity to investigate biomolecular interaction by implementing fuzzy logic sets as proposed by Gentili [36].

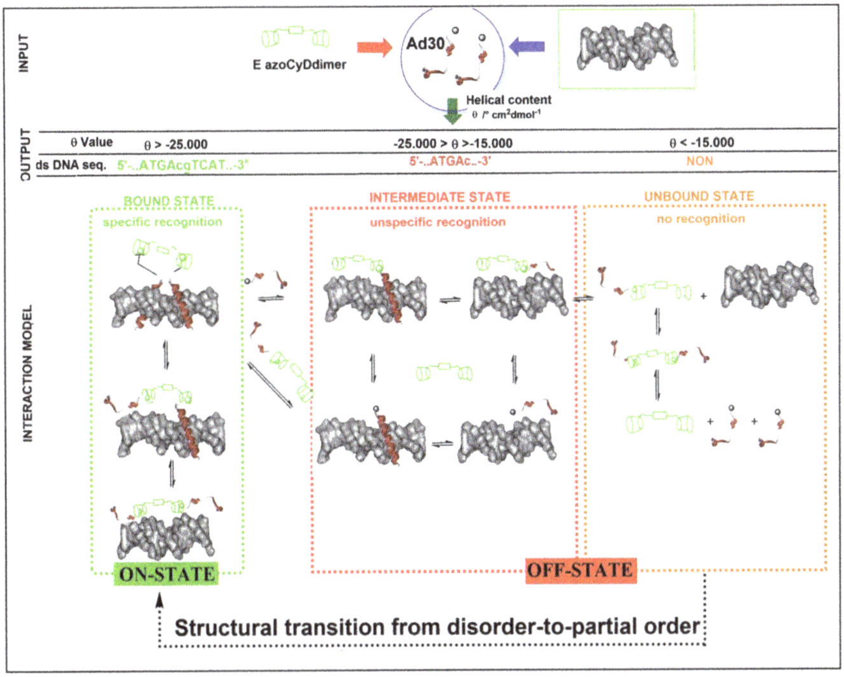

Figure 8. Schematic structural transitions of the Ad30 upon partner interaction with AzoCyDdimer (E) in the presence of ds CRE, representing the interacting molecules (INPUT), the obtained experimental values (OUTPUT), and the proposed interaction model. Three interaction stages are hypothesized for the system: the unbound-, the intermediate-, and the bound-state that are represented with dashed rectangles in orange, red, and green, respectively. These stages are based on the experimental helical content (θ) obtained from the individual experiments in the presence of CRE, AP1, mCRE, or NON, and the stoichiometry obtained in the EMSA experiments. In each stage, different complexes and their conformational fuzziness are shown. In the on-state *(bound state, green dashed rectangle)*, the specific binding to CRE builds up the tetra-component complex. In the bound state, the structural and dynamical continuum of fuzzy complexes is shown. The off-state is composed of many complexes, which can be unbounded to the dsDNA (*orange dashed rectangle)* or bounded but without the required sequence-specificity (*red dashed rectangle)*, thus they are referred here as the intermediate state. The structural transition forms the unbound state (disorder) to the bound state (partial order) is shown.

4. Materials and Methods

4.1. Peptide Synthesis

Disulfide dimers SS52 and SS60 were synthesized from commercial peptides SH26 and SH30, to have reference standards to study their interactions with DNA [15]. Both products were purified by semi-preparative reverse phase HPLC (Waters, Milford, MA, USA) and then lyophilized. The characterization was carried out by MALDI–TOF (Bruker Daltonik, Bremen, Germany), SS52: (M + H) calc. $C_{234}H_{414}N_{98}O_{74}S_2$ 5786, found: 5788.5 (43%yield). SS60: $(M + H)^+$ calc. $C_{278}H_{498}N_{116}O_{80}S_4$ 6873.95, found: 6875.36 (50% yield).

For the synthesis of Ad26 [15] and Ad30, on both deoxygenated solutions of SH26 (1.9 mg, 6.2×10^{-4} mmol) and SH30 (2.0 mg, 5.82×10^{-4} mmol) in potassium phosphate buffer (150 μL, 100 mM, pH = 8.0) and CH_3CN 50 μL were added 4 equiv of bromo acetyldamantane in CH_3CN (0.79 mg, 9 μL). The mixtures were stirred at room temperature for 3 h under N_2 and checked by RP-LC–MS (Agilent 1100, Santa Clara, CA, USA),. The new compounds were purified by semi-preparative RP-HPLC and then lyophilized, identified by mass spectrometry as the alkylated peptides. Once purified and lyophilized, Ad26 (59%) and Ad30 were obtained in 59% and 66% yield, respectively. MALDI–TOF for Ad26, $C_{130}H_{229}N_{49}O_{37}S_1$: calc. $(M + H)^+$ 3101.36, found 3102.81; and for Ad30, $C_{152}H_{272}N_{58}O_{42}S_2$: calc. $(M + H)^+$ 3646.04, found 3647.08.

4.2. Synthesis and Characterization of azoCyDdimer

4,4′-bis (carboxy) azobenzene (0.036 g, 0.132 mmol) was dissolved in dry DMF (2 mL), HATU in DMF was added (0.10 g, 0.317 mmol) and 0.183 mL DIPEA (0.136 g, 1.056 mmol), was stirred for 5 min at room temperature, and to this solution was added β-CyD-NH2 (0.300 g, 0.264 mmol) dissolved in 2 mL of DMF, the resulting mixture was stirred for 3 h under nitrogen atmosphere. The mixture was poured over a container with ice-cold acetone and a precipitate appeared. The solid was separated by filtration and then dissolved in DMF to be purified by silica gel chromatography column. (CH_3CN–H_2O–NH_4OH 14:0:0.5/4:10:0.5). ^{1}H-NMR (500 MHz (Avance II 500, Bruker, Germany), DMSO-d6) δ (ppm): 3.3–4.33 (m, 42H ov 2 H_2O), 4.33–4.56 (br.s, 12H), 4.96 (br.s, 14H), 4.82 (br.s, 7H), 5.7–5.82 (m, OH, 27H), 7.98 (4 H, d, J = 8.6 Hz), 8.04 (4 H, d, J = 8.6 Hz), 8.50 (2 H, br. s). ^{13}C-NMR (125.75 MHz, DMSO-d_6) δ (ppm): 41.8 (CH_2), 59.7 (CH_2), 71.9 (CH), 72.3 (CH), 73.0 (CH), 81.3 (CH), 81.9 (CH), 84.1 (CH), 101.64 (CH), 101.9 (CH), 122.2 (CH), 128.4 (CH), 137.4 (C), 153.7 (C), 166.2 (C). MALDI–TOF $(M + H)^+$: calcd. for $C_{98}H_{148}N_4O_{70}$ 2500.8, $(M + Na)^+$: calc. 2523.58, found 2523.53, $(M + K)^+$: calc. 2539.98, found 2539.96.

H-NMR Photoisomerization Experiment

Four independent photoisomerization experiments were performed using different initial concentrations of azoCyDdimer: 15 mM, 8 mM, 2.28 mM, and 0.5 mM. Dilutions were made from a 15 mM stock solution with an isomer ratio of 95:5 (E:Z). The stock solution and the dilutions were irradiated at 360 nm for 20 min, and then the corresponding ^{1}HRMN spectra were acquired (AMX 300, Bruker, Germany), protecting the solution from the visible light. Switching experiments were performed with an 8 W mercury arc lamp with filter of 360 nm from Pleuger, Antwerp, Brussels.

4.3. Annealing of dsDNA

Oligonucleotides were purchased from Thermo Fisher Scientific GmbH on a 0.2 mmol scale as freeze-dried solids. After solving in H_2O milliQ, their concentrations were measured by ultraviolet absorption at 260 nm with a BioRad SmartSpec Plus Spectrophotometer. Absorbance was measured twice, and concentrations were calculated applying Lambert–Beer's equation. The molar extinction coefficients of single-strand oligonucleotides were calculated by using the following formula ε(260 nm) = $\{(8.8 \times \#T) + (7.3 \times \#C) + (11.7 \times \#G) + (15.4 \times \#A)\} \times 0.9 \times 10^3$ M^{-1} cm^{-1}, where #A, #T, #C, #G stand for the number of each type of bases in the DNA strand. Oligonucleotides were annealed by heating an equimolar mixture of the two complementary single-strand DNAs to 90 °C for 2 min and then cooling slowly to room temperature (12 h).

4.4. Electrophoretic Shift binding Assays

EMSA was performed with a BioRad Mini Protean gel system, powered by an electrophoresis power supplies PowerPac Basic model, maximum power 150 V, frequency 50.60 Hz at 140 V (constant V). The binding reactions were performed over 30 min in a binding mixture (20 or 40 μL) containing 18 mm tris(hydroxymethyl)aminomethane (Tris; pH 7.5), 90 mm KCl, 1.8 mm $MgCl_2$, 1.8 mm EDTA, 9%glycerol, 0.11 $mgmL^{-1}$ bovine serum albumin (BSA), and 2.2% NP-40 (nonidet-P40). Products were resolved by PAGE by using a 10% nondenaturing polyacrylamide gel and 0.5XTBE buffer solution (44.5 mm Tris, 44.5 mm boric acid,1 mm EDTA, pH 8) and analyzed by staining with SyBrGold (Molecular Probes: 5 mL in 50 mL of 1XTBE) for 10 min and visualized with fluorescence. Ad26 working concentration was 200 nM, 50 nM of DNA, and for azoCyDdimer was used from 0 to 100 equivalents of the E and Z dimer, respectively, was selected (stock solution 0.74 mM in H_2O, ratio E:Z determined by RP-HPLC, Supplementary Materials Figure S5). The order of addition was azoCyDdimer, Ad26 (pre-incubation for 10 min at 4 °C), and then the corresponding DNA. Duplicates of independent experiments were performed. For the competitive binding assay, we added 200 equiv of 1-adamantane acetic acid after the addition of azoCyDdimer.

4.5. Circular Dichroism Spectroscopy

CD experiments were performed on a Jasco spectrometer (Jasco 715, Tokyo, Japan) with 1 mm path-length cell [17]. Samples in 10 mM phosphate buffer (pH 7.0) and 100 mM NaCl contained 10 or 5 µM of peptide and 5 µM oligonucleotides (double-stranded) in the absence or presence of 1 equiv of azoCyDdimer at 4 °C (stock solution 0.74 mM in H_2O, ratio E:Z determined by RP-HPLC, see Supplementary Materials Figure S5). Every sample was incubated for 5 min before registering. Duplicates of independent experiments were performed. The CD spectra of the peptides (when measured in the presence of DNA) were calculated as the difference between the spectrum of the peptide/DNA mixture and the measured spectrum of the respective dsDNA oligonucleotide. For the competitive binding assay, we added 200 equiv of 1-adamantane acetic acid after the addition of azoCyDdimer. Smoothing of the signals was performed by using the software KaleidaGraph.

Supplementary Materials: The following are available online at http://www.mdpi.com/1420-3049/24/13/2508/s1, Figure S1: ^{1}H-NMR of the azoCyDdimer, Figure S2: 2D-NMR-HSQC of azoCyDdimer, Figure S3: 2D-NMR-HMBC of azoCyDdimer, Figure S4: UV–Vis Spectra of azoCyDdimer depending on illumination conditions; Figure S5 RP-LC–MS of azoCyDdimer depending on illumination conditions. Tables S1 and S2: ^{1}H-NMR and ^{13}C-NMR chemical shifts of AzoCyDdimer (E).

Author Contributions: Conceptualization, V.I.D.; Formal analysis, Z.B.Q. and V.I.D.; Funding acquisition, Z.B.Q. and V.I.D.; Investigation, Z.B.Q. and M.A.S.; Methodology, Z.B.Q., J.C.M., and V.I.D.; Resources, J.C.M. and V.I.D.; Supervision, J.C.M. and V.I.D.; Writing—original draft, Z.B.Q.; Writing—review and editing, Z.B.Q., M.A.S., J.C.M., and V.I.D.

Funding: This research was funded by National Council of Scientific and Technical Research of Argentina (CONICET, PIP 114-200801-00090), National University of South grant PGI-UNS 2014, MINCYT-DAAD program with University of Bielefeld. The APC was funded by the Deutsche Forschungsgemeinschaft and the Open Access Publication Fund of Bielefeld University.

Acknowledgments: Z.B.Q. is very grateful for her CONICET fellowships, Erasmus Mundus fellowship (EuroTango) in J. C. Martins' group and Santander-USC fellowship in J. L. Mascareñas' group. M.A.S. thanks her CONICET fellowship. V.I.D. thanks N. Sewald and J. L. Mascareñas for their support.

Conflicts of Interest: The authors declare no conflict of interest. The funders had no role in the design of the study; in the collection, analyses, or interpretation of data; in the writing of the manuscript; or in the decision to publish the results.

References

1. Tompa, P.; Fuxreiter, M. Fuzzy complexes: Polymorphism and structural disorder in protein-protein interactions. *Trends Biochem. Sci.* **2008**, *33*, 2–8. [CrossRef] [PubMed]
2. Sharma, R.; Raduly, Z.; Miskei, M.; Fuxreiter, M. Fuzzy complexes: Specific binding without complete folding. *FEBS Lett.* **2015**, *589*, 2533–2542. [CrossRef] [PubMed]
3. Gentili, P.L. The Fuzziness of the Molecular World and its Perspectives. *Molecules* **2018**, *23*, 2074. [CrossRef]
4. Gentili, P.L. *Untangling Complex Systems: A Grand Challenge for Science*, 1st ed.; CRC Press: Boca Raton, FL, USA, 2018.
5. Arbesú, M.; Iruela, G.; Fuentes, H.; Teixeira, J.M.C.; Pons, M. Intramolecular Fuzzy Interactions Involving Intrinsically Disordered Domains. *Front. Mol. Biosci.* **2018**, *5*, 39. [CrossRef] [PubMed]

6. Mösch, H.U.; Reusser, U.; Albrecht, G.; Braus, G.H.; Hoffmann, B. Monitoring the Gcn4 Protein-mediated Response in the Yeast Saccharomyces cerevisiae. *J. Biol. Chem.* **2002**, *273*, 12696–12702.
7. Ellenberger, T.E.; Brandl, C.J.; Struhl, K.; Harrison, S.C. The GCN4 basic region leucine zipper binds DNA as a dimer of uninterrupted α Helices: Crystal structure of the protein-DNA complex. *Cell* **1992**, *71*, 1223–1237. [CrossRef]
8. König, P.; Richmond, T.J. The X-ray structure of the GCN4-bZIP bound to ATF/CREB site DNA shows the complex depends on DNA flexibility. *J. Mol. Biol.* **1993**, *233*, 139–154. [CrossRef]
9. Tuttle, L.M.; Pacheco, D.; Warfield, L.; Luo, J.; Ranish, J.; Hahn, S.; Klevit, R.E. Gcn4-Mediator Specificity is Mediated by a Large and Dynamic Fuzzy Protein-Protein Complex. *Cell Rep.* **2018**, *22*, 3251–3264. [CrossRef]
10. Talanian, R.V.; Mcknight, C.J.; Kim, P.S. Sequence-specific DNA binding by a short peptide dimer. *Science* **1990**, *249*, 769–771. [CrossRef]
11. Talanian, R.V.; McKnight, C.J.; Rutkowski, R.; Kim, P.S. Minimum Length of a Sequence-Specific DNA Binding Peptide. *Biochemistry* **1992**, *31*, 6871–6875. [CrossRef]
12. Morii, T.; Yamane, J.; Aizawa, Y.; Makino, K.; Sugiura, Y. Cooperative Oligomerization Enhances Sequence-Selective DNA Binding by a Short Peptide. *J. Am. Chem. Soc.* **1996**, *118*, 10011–10017. [CrossRef]
13. Caamaño, A.; Vázquez, M.; Martínez-Costas, J.; Castedo, L.; Mascareñas, J.L. A Light-Modulated Sequence-Specific DNA-Binding Peptide. *Angew. Chem. Int. Ed. Engl.* **2000**, *39*, 3104–3107. [CrossRef]
14. Mosquera, J.; Jiménez-Balsa, A.; Dodero, V.I.; Vázquez, M.E.; Mascareñas, J.L. Stimuli-Responsive Selection of Target DNA Sequences by Synthetic bZIP Peptides. *Nat. Commun.* **2013**, *4*, 1874. [CrossRef] [PubMed]
15. Breslow, R.; Greenspoon, N.; Guo, T.; Zarzycki, R. Very strong binding of appropriate substrates by cyclodextrin dimers. *J. Am. Chem. Soc.* **1989**, *111*, 8296–8297. [CrossRef]
16. Hamon, F.; Blaszkiewicz, C.; Buchotte, M.; Banaszak-Léonard, E.; Bricout, H.; Tilloy, S.; Monflier, E.; Cézard, C.; Bouteiller, L.; Len, C.; et al. Synthesis and Characterization of a New Photoinduced Switchable β-Cyclodextrin Dimer. *Beilstein J. Org. Chem.* **2014**, *10*, 2874–2885. [CrossRef] [PubMed]
17. Blanco, J.B.; Dodero, V.I.; Vazquez, M.E.; Mosquera, M.; Castedo, L.; Mascareñas, J.L. Sequence-Specific DNA Binding by Noncovalent Peptide-Tripyrrole Conjugates. *Angew. Chem.* **2006**, *118*, 8390–8394. [CrossRef]
18. Ueno, M.; Murakami, A.; Makino, K. Cyclodextrin-Guest Inclusion Complex: Sequence-Specific DNA Binding by a Peptide Dimer with Artificial Dimerization Module. *Communications* **1993**, *115*, 12575–12576.
19. Hellman, L.M.; Fried, M.G. Electrophoretic Mobility Shift Assay (EMSA) for detecting protein-nucleic acid interactions. *Nat. Protoc.* **2007**, *2*, 1849–1861. [CrossRef] [PubMed]
20. Bandara, H.M.D.; Burdette, S.C. Photoisomerization in different classes of azobenzene. *Chem. Soc. Rev.* **2012**, *41*, 1809–1825. [CrossRef]
21. Sequeira, M.A.; Herrera, M.G.; Quirolo, Z.B.; Dodero, V.I. Easy directed assembly of only nonionic azoamphiphile builds up functional azovesicles. *RSC Adv.* **2016**, *6*, 108132–108135. [CrossRef]
22. Zhu, M.; Zhou, H. Azobenzene-Based Small Molecular Photoswitches for Protein Modulation. *Org. Biomol. Chem.* **2018**, *16*, 8434–8445. [CrossRef] [PubMed]
23. Kelly, S.M.; Jess, T.J.; Price, N.C. How to study proteins by circular dichroism. *Biochim. Biophys. Acta-Proteins Proteom.* **2005**, *1751*, 119–139. [CrossRef] [PubMed]
24. Dodero, V.I.; Quirolo, Z.B.; Sequeira, M.A. Biomolecular studies by circular dichroism. *Front. Biosci.* **2011**, *16*, 61–73. [CrossRef]
25. Fuxreiter, M.; Tompa, P. Fuzzy complexes: A more stochastic view of protein function. *Adv. Exp. Med. Biol.* **2012**, *725*, 1–14. [PubMed]
26. Wright, P.E.; Dyson, H.J. Intrinsically unstructured proteins: Re-assessing the protein structure-function paradigm. *J. Mol. Biol.* **1999**, *293*, 321–331. [CrossRef] [PubMed]
27. Fuxreiter, M. Fuzziness in Protein Interactions—A Historical Perspective. *J. Mol. Biol.* **2018**, *430*, 2278–2287. [CrossRef] [PubMed]
28. Fuxreiter, M. Fold or not to fold upon binding—does it really matter? *Curr. Opin. Struct. Biol.* **2018**, *54*, 19–25. [CrossRef]
29. Pricer, R.; Gestwicki, J.E.; Mapp, A.K. From fuzzy to function: The new frontier of protein-protein interactions. *Acc. Chem. Res.* **2017**, *50*, 584–589. [CrossRef]
30. Cranz, S.; Berger, C.; Baici, A.; Jelesarov, I.; Bosshard, H.R. Monomeric and Dimeric bZIP Transcription Factor GCN4 Bind at the Same Rate to Their Target DNA Site. *Biochemistry* **2004**, *43*, 718–727. [CrossRef]

31. Iyer, A.; Garcia, Y.R.; Madder, A. *Synthetic Peptides for DNA Recognition Inspired by Transcription Factors*; RSC: London, UK, 2018; Chapter 13.
32. Miskei, M.; Gregus, A.; Sharma, R.; Duro, N.; Zsolyomi, F.; Fuxreiter, M. Fuzziness enables context dependence protein interactions. *FEBS Lett.* **2017**, *591*, 2682–2695. [CrossRef]
33. Del Sol, A.; Tsai, C.J.; Ma, B.; Nussinov, R. The origin of allosteric functional modulation: Multiple pre-existing pathways. *Structure* **2009**, *17*, 1042–1050. [CrossRef] [PubMed]
34. Wright, P.E.; Dyson, H.J. Linking folding and binding. *Curr-Opin. Struct. Biol.* **2009**, *19*, 31–38. [CrossRef] [PubMed]
35. Kitov, P.I.; Bundle, D.R. On the nature of the multivalency effect: A thermodynamic model. *J. Am. Chem. Soc.* **2003**, *125*, 16271–16284. [CrossRef] [PubMed]
36. Gentili, P.L. Boolean and fuzzy logic implemented at the molecular level. *Chem. Phys.* **2007**, *336*, 64–73. [CrossRef]

Sample Availability: Samples of the compounds are not available from the authors.

molecules

MDPI

Article

From the Kinetic Theory of Gases to the Kinetics of Rate Processes: On the Verge of the Thermodynamic and Kinetic Limits

Valter H. Carvalho-Silva [1,*], Nayara D. Coutinho [2,*] and Vincenzo Aquilanti [3,4,*]

1 Divisão de Modelagem de Transformações Físicas e Químicas, Grupo de Química Teórica e Estrutural de Anápolis, Centro de Pesquisa e Pós-Graduação. Universidade Estadual de Goiás, Anápolis 75132-903, Brazil
2 Instituto de Química, Universidade de Brasília, Caixa Postal 4478, Brasília 70904-970, Brazil
3 Dipartimento di Chimica, Biologia e Biotecnologie, Università di Perugia, 06123 Perugia, Italy
4 Istituto di Struttura della Materia, Consiglio Nazionale delle Ricerche, 00133 Rome, Italy
* Correspondence: fatioleg@gmail.com (V.H.C.-S.); nayaradcoutinho@gmail.com (N.D.C.); vincenzo.aquilanti@unipg.it (V.A.); Tel.: +55-62-3328-1156 (V.H.C.-S.)

Academic Editor: Pier Luigi Gentilli
Received: 8 April 2020; Accepted: 28 April 2020; Published: 30 April 2020

Abstract: A variety of current experiments and molecular dynamics computations are expanding our understanding of rate processes occurring in extreme environments, especially at low temperatures, where deviations from linearity of Arrhenius plots are revealed. The thermodynamic behavior of molecular systems is determined at a specific temperature within conditions on large volume and number of particles at a given density (the thermodynamic limit): on the other side, kinetic features are intuitively perceived as defined in a range between the extreme temperatures, which limit the existence of each specific phase. In this paper, extending the statistical mechanics approach due to Fowler and collaborators, ensembles and partition functions are defined to evaluate initial state averages and activation energies involved in the kinetics of rate processes. A key step is delayed access to the thermodynamic limit when conditions on a large volume and number of particles are not fulfilled: the involved mathematical analysis requires consideration of the role of the succession for the exponential function due to Euler, precursor to the Poisson and Boltzmann classical distributions, recently discussed. Arguments are presented to demonstrate that a universal feature emerges: Convex Arrhenius plots (*super*-Arrhenius behavior) as temperature decreases are amply documented in progressively wider contexts, such as viscosity and glass transitions, biological processes, enzymatic catalysis, plasma catalysis, geochemical fluidity, and chemical reactions involving collective phenomena. The treatment expands the classical Tolman's theorem formulated quantally by Fowler and Guggenheim: the activation energy of processes is related to the averages of microscopic energies. We previously introduced the concept of "transitivity", a function that compactly accounts for the development of heuristic formulas and suggests the search for universal behavior. The velocity distribution function far from the thermodynamic limit is illustrated; the fraction of molecules with energy in excess of a certain threshold for the description of the kinetics of low-temperature transitions and of non-equilibrium reaction rates is derived. Uniform extension beyond the classical case to include quantum tunneling (leading to the concavity of plots, *sub*-Arrhenius behavior) and to Fermi and Bose statistics has been considered elsewhere. A companion paper presents a computational code permitting applications to a variety of phenomena and provides further examples.

Keywords: Maxwell–Boltzmann path; Euler's formula for the exponential; activation; transitivity; transport phenomena

1. Introduction

A basic task of current molecular science is to elucidate how the kinetic behavior of a physicochemical system manifests within the temperature range of its "life span": thermodynamics has its focus on states of the system and the transition between them, while the study of the rate of evolution of processes is the objective of kinetics. In thermodynamics (as in mechanics), it is ubiquitous to face the balance among various types of energy being exchanged; the connection from the molecular to macroscopic energy levels requires averages over the myriads of ways of change of microscopic configurations that determine the progress of events. The situation in chemical kinetics is intrinsically not so sharp, not only because systems in movement are much harder to be studied than those in steady-state equilibrium. Currently, the techniques in experimental and theoretical kinetics have been advancing enormously (although at a much slower pace when compared with those of thermodynamics), due to progress on production and detection of molecular beams and on classical and quantum simulations of molecular dynamics.

Aspects related to the foundations of the kinetics of rate processes were elaborated recently in previous papers [1–4]. In [1], fundamental concepts concerning statistical distributions and reaction rate theory were presented, including the definition of transitivity, a function of absolute temperature denoted as $\gamma(T)$, based on extensive phenomenology that is being accumulated; a subsequent paper [2] considered the historical background of developments of chemical kinetics, leading to the basic foundations through analysis of key mathematical ingredients; in [3], the formulations based on the concept of transitivity were compacted and applied to the description of several phenomena on the temperature dependence of rate processes beyond Arrhenius and Eyring; and finally in paper [4], companion of this one in this topical issue, a computational code is described and provided to calculate kinetics and related parameters in chemical transformations and transport phenomena.

The need emerges of differentiating developments from those employed in thermodynamics, in spite of the fact that the kinetic theory of gases by Maxwell (and later by Boltzmann) was formulated more or less contemporary to the thermodynamics of Carnot (and later of Clausius): their thermodynamic vision was later merged turning the Maxwell theory essentially in terms of Boltzmann–Gibbs distributions. Additionally, as a well-known matter of fact in the literature [5–7], the Arrhenius equation, basic to chemical kinetics, was suggested as an empirical adaptation of the thermodynamics of chemical equilibrium developed by van't Hoff (ca. 1880).

In the present work, account will be given to how a derivation of a theory of rate processes from non-equilibrium distributions involves essentially steps that are usual in thermodynamics, specifically as far as averaging procedures are concerned. A specific feature here is that we are progressing in the spirit of the well-known Darwin–Fowler formulation [8–10], which involves departure from the concept of "most probable" configuration emerging following the Boltzmann–Gibbs path [11,12]. Darwin and Fowler dealt with average quantities: they essentially developed a thermodynamics equivalent to the canonical form with no need of the concept of a microcanonical ensemble or even of that of entropy: a similar alternative path was briefly indicated by Eyring and coworkers presenting the foundations of the "Transition-State Theory" of rate processes [13–16]. This approach appears better motivated than the traditional: current experiments involve molecular beams studies of individual events, and advances in quantum mechanical treatments indicate the "royal path": theoretical chemical kinetics proceeds by generating intermolecular potential energy surfaces and simulate computationally the passage from myriads of microscopic events to macroscopic quantities. This can be formulated at least in principle: however, when we consider polyatomic systems in molecular dynamics, we are unable to fully circumventing the difficulty presented by the need of averaging on a large number of events, difficult to be sampled in a statistically converged way. The situation was first anticipated by Maxwell who conjectured that the collective macroscopic observable motion of atoms, if they existed, should be compacted by averaging over statistical ensembles of their "trajectories" [17,18].

Following downward the upper part of the chart in Figure 1, we consider ab initio 'exact' quantum dynamics: it is expected to provide benchmark kinetics data, but is in practice still limited to simple cases, see [19–21]. Such applications of exemplary chemical reaction kinetics typically proceed according to the descending steps schematized in Figure 1: (a) calculation of the molecular electronic structure interactions involving high-level of quantum chemical accuracy, (b) dynamical evolution in phase-space configurations from the solution of the (usually time-independent) quantum equations of the motion, and (c) extraction of reactive properties from asymptotic scattering theory and calculating in succession key quantities: the quantum scattering matrix, the cumulative reaction probability and the cross sections. Finally, the Boltzmann weight averaging over a large span and fine grid of kinetic energies is needed to obtain the canonical quantity of chemical kinetics, the rate constants $k(\mathrm{T})$ as a function of temperature. These rigorous prescriptions can yield benchmark results for quantum evolution of systems over a given potential energy surface and provide reaction rate constants for only a limited number of reactions: in fact, the complexity of the programming and the computationally demanding requirements strongly limit this type of study to reactive processes involving only a few atoms [19,21,22]. However, they serve as prototypes to complex molecular systems and stepping stones for example to processes governed by multiple potential energy surfaces, nonadiabatically coupled.

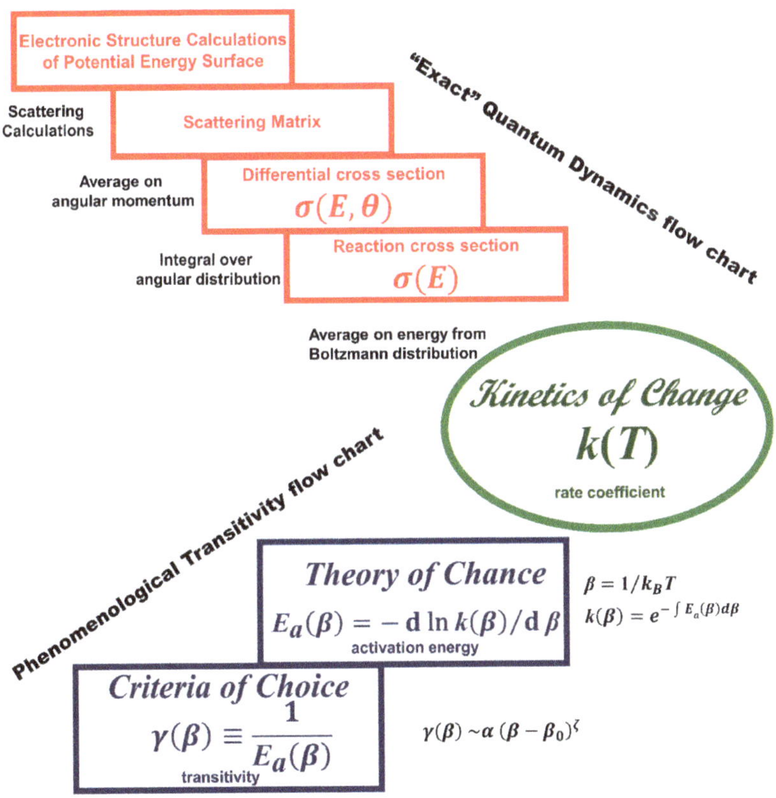

Figure 1. Stepping stones for the sequence (in red) from quantum chemical calculations to theoretical ("exact") rate coefficients (downward along upper part of the chart, see also [2]); and the sequence (in blue) from the transitivity function $\gamma(\beta)$ to the phenomenological rate coefficients (upward in the lower part of the chart). Models for $\gamma(\beta)$ are discussed in [3] and further in Sections 3 and 4. Here $\alpha = 1/\varepsilon^{\ddagger}$, the reciprocal of Eyring activation energy for $\beta_0 = 0$ in the asymptotically high temperature of the thermodynamic limit (Section 2.2). The critical exponent ξ generalizes the integer n in reference [3] and will be related (see Figure 5) to classes of universal behavior.

In this paper, the phenomenological rate theory [1–3] is developed by introducing a mechanism where the delay or acceleration of the approach to a well-defined mathematical limit due to Euler accounts for the low-temperature deviations of rates from Arrhenius law. In the next section, revisitation of the classical thermodynamic limit accompanies its extension to kinetics and naturally leads to a deformation of the Boltzmann–Gibbs distribution and to the emergence of a formulation alternative to that of Arrhenius [1–3]. Section 3 illustrates how theory serves to the natural scaling of a variety of physical and chemical processes in extreme conditions and near phase transitions. Implementations to various phenomena are sketched in Section 4. Concluding remarks are given in Section 5. Appendix A presents formulas for the distribution of energies in a reactive process away from equilibrium.

2. Thermodynamic versus Kinetic Limits, Revisited

2.1. *The Exponential as Limit of Euler's Succession: Role in the Early Kinetic Theory of Gases*

The memorable succession for the exponential function from the sum of an infinite series is a powerful variant of the binomial theorem of Newton and was discovered by Euler in the XVIII century. For its occurrence originated in a famous bank account problem solved by Jacob Bernoulli and for aspects of its remarkable facets, see the recent papers [1–3,23]. The tremendous advances in the kinetic theory of gases started in the mid-XIX century with Maxwell's mathematically intuition to look at the microscopic world as composed of greatly many indivisible particles, atoms. According to this vision, which found in Ludwig Boltzmann [24] one of the greatest defenders in times when even the existence of atoms was being questioned, the germs of what is now known as statistical mechanics were formulated: the motion of microscopic particles was correlated to macroscopic observables providing the foundations for the phenomenology of thermodynamics. It is not always recognized that the statistical proposition for success or failure of events (Bernoulli urn or Bernoulli trial binomial process and its generalization) provided through the Euler's limit the foundations for the derivation of extensions to distributions, i.e., the foundations for the progress in the XIX and the early XX century, remarkably those of Poisson, Gauss, Planck, Bose–Einstein, and Fermi–Dirac [1].

At the very origin of the statistical mechanics viewpoint, the investigations reported in the 1860 [17] and 1866 [18] papers by James Clerk Maxwell lead to the famous velocity distribution of molecules under the hypothesis of the independence of Cartesian components of the velocity vectors: this conjecture appeared plausible from the additive properties of the exponential function. In 1868, Ludwig Boltzmann [25], as reviewed, e.g., in reference [26], introduced probabilistic concepts—the "marginal" probability of the energy of a molecule—obtaining a derivation of the Maxwell's law of velocities by rigorous treatment based explicitly on the exponential behavior of velocity according to the Euler's succession, see Figure 2.

A decade after, Maxwell [27] returns to the Boltzmann's formulation proposing a more insightful approach of the problem, see Figure 3, rarely considered in the expositions of the theory in wide number of papers, treatises, and textbooks. In 1916, Jeans in his treatise on "The Dynamical Theory of Gases" [28] addresses Maxwell's latest treatment within a much more concise mathematical analysis generalizing the concept of phase-space: again the exponential velocity distribution law is obtained from the Euler's limit of a succession. This same procedure can be traced in further [29] and recent [30] works, where cases involving finite systems are dealt with essentially by arresting the treatment before taking Euler's limit, namely without taking what is now recognized as the "thermodynamic limit".

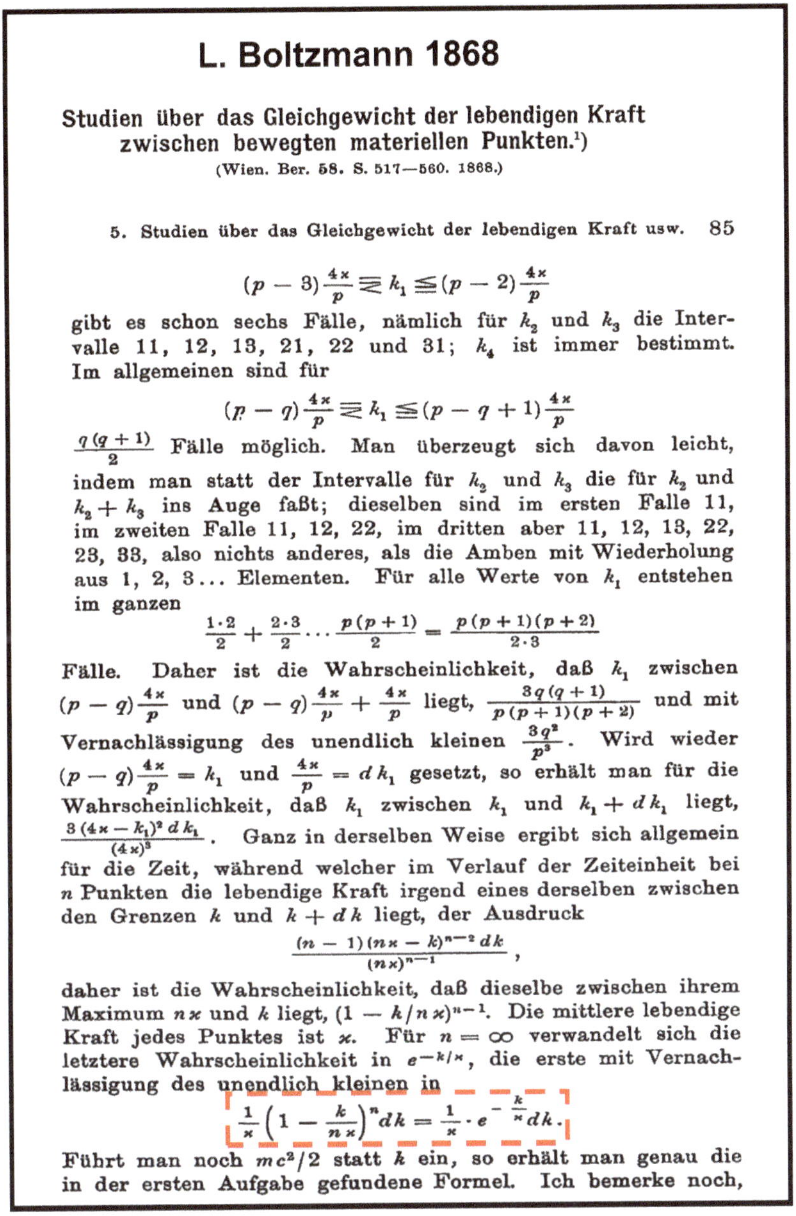

L. Boltzmann 1868

Studien über das Gleichgewicht der lebendigen Kraft zwischen bewegten materiellen Punkten.[1])

(Wien. Ber. 58. S. 517—560. 1868.)

5. Studien über das Gleichgewicht der lebendigen Kraft usw. 85

$$(p-3)\frac{4\varkappa}{p} \gtreqless k_1 \leqq (p-2)\frac{4\varkappa}{p}$$

gibt es schon sechs Fälle, nämlich für k_2 und k_3 die Intervalle 11, 12, 13, 21, 22 und 31; k_4 ist immer bestimmt. Im allgemeinen sind für

$$(p-q)\frac{4\varkappa}{p} \gtreqless k_1 \leqq (p-q+1)\frac{4\varkappa}{p}$$

$\frac{q(q+1)}{2}$ Fälle möglich. Man überzeugt sich davon leicht, indem man statt der Intervalle für k_2 und k_3 die für k_2 und $k_2 + k_3$ ins Auge faßt; dieselben sind im ersten Falle 11, im zweiten Falle 11, 12, 22, im dritten aber 11, 12, 13, 22, 23, 33, also nichts anderes, als die Amben mit Wiederholung aus 1, 2, 3... Elementen. Für alle Werte von k_1 entstehen im ganzen

$$\frac{1\cdot 2}{2} + \frac{2\cdot 3}{2} \cdots \frac{p(p+1)}{2} = \frac{p(p+1)(p+2)}{2\cdot 3}$$

Fälle. Daher ist die Wahrscheinlichkeit, daß k_1 zwischen $(p-q)\frac{4\varkappa}{p}$ und $(p-q)\frac{4\varkappa}{p} + \frac{4\varkappa}{p}$ liegt, $\frac{3q(q+1)}{p(p+1)(p+2)}$ und mit Vernachlässigung des unendlich kleinen $\frac{3q^2}{p^3}$. Wird wieder $(p-q)\frac{4\varkappa}{p} = k_1$ und $\frac{4\varkappa}{p} = dk_1$ gesetzt, so erhält man für die Wahrscheinlichkeit, daß k_1 zwischen k_1 und $k_1 + dk_1$ liegt, $\frac{3(4\varkappa - k_1)^2\,dk_1}{(4\varkappa)^3}$. Ganz in derselben Weise ergibt sich allgemein für die Zeit, während welcher im Verlauf der Zeiteinheit bei n Punkten die lebendige Kraft irgend eines derselben zwischen den Grenzen k und $k + dk$ liegt, der Ausdruck

$$\frac{(n-1)(n\varkappa - k)^{n-2}\,dk}{(n\varkappa)^{n-1}},$$

daher ist die Wahrscheinlichkeit, daß dieselbe zwischen ihrem Maximum $n\varkappa$ und k liegt, $(1 - k/n\varkappa)^{n-1}$. Die mittlere lebendige Kraft jedes Punktes ist $\varkappa$. Für $n = \infty$ verwandelt sich die letztere Wahrscheinlichkeit in $e^{-k/\varkappa}$, die erste mit Vernachlässigung des unendlich kleinen in

$$\frac{1}{\varkappa}\left(1 - \frac{k}{n\varkappa}\right)^n dk = \frac{1}{\varkappa}\cdot e^{-\frac{k}{\varkappa}}\,dk.$$

Führt man noch $mc^2/2$ statt k ein, so erhält man genau die in der ersten Aufgabe gefundene Formel. Ich bemerke noch,

Figure 2. The page of the Boltzmann's paper where he used a combinatorial argument that we follow in the formulation exposed in the Appendix A. He obtained the Maxwell distribution through the Euler formula (boxed) giving the exponential function as a limit of a succession.

J.C. Maxwell 1879

[From the *Cambridge Philosophical Society's Transactions*, Vol. XII.]

XCIV. *On Boltzmann's Theorem on the average distribution of energy in a system of material points.*

726 BOLTZMANN'S THEOREM ON THE AVERAGE DISTRIBUTION

The whole number of systems in configuration (b) is given by (45). Hence the proportion of these systems for which a_n lies between a_n and $a_n + da_n$ is

$$\frac{2^{-\frac{1}{2}}\Gamma\left(\frac{n}{2}\right)}{\Gamma\left(\frac{1}{2}\right)\Gamma\left(\frac{n-1}{2}\right)} \frac{\left[E-V-\frac{1}{2}\mu_n a_n^2\right]^{\frac{n-3}{2}}}{\left[E-V\right]^{\frac{n-2}{2}}}\mu_n^{\frac{1}{2}}\,da_n \ldots\ldots\ldots\ldots\ldots\ldots(49).$$

If we write

$$\tfrac{1}{2}\mu_n a_n^2 = k_n \ldots\ldots\ldots\ldots\ldots\ldots\ldots\ldots\ldots\ldots(50),$$

then k_n denotes the part of the kinetic energy arising from the momentum a_n. The proportion of the systems in configuration (b) for which k_n is between k_n and $k_n + dk_n$ is

$$\frac{\Gamma\left(\frac{n}{2}\right)}{\Gamma\left(\frac{1}{2}\right)\Gamma\left(\frac{n-1}{2}\right)} \frac{\left[E-V-k_n\right]^{\frac{n-3}{2}}}{\left[E-V\right]^{\frac{n-2}{2}}}k_n^{-\frac{1}{2}}\,dk_n \ldots\ldots\ldots\ldots\ldots\ldots(51).$$

Since any one of the variables may be taken for q_n, the law of distribution of values of the kinetic energy is the same for all the variables. The mean value of the kinetic energy corresponding to any variable is

$$K = \frac{1}{n}(E-V) = \frac{1}{n}T \ldots\ldots\ldots\ldots\ldots\ldots\ldots\ldots(52).$$

The maximum value is $T = nK \ldots\ldots\ldots\ldots\ldots\ldots\ldots\ldots\ldots\ldots(53)$.

The mean value of k^r is

$$\frac{1\,.\,3\ldots 2r-1}{n\,.\,n+2\ldots n+2r-2}n^r K^r \ldots\ldots\ldots\ldots\ldots\ldots\ldots\ldots(54).$$

When n is very large, the expression (51) approximates to

$$\frac{1}{\sqrt{2\pi}}\frac{1}{K}e^{-\frac{k}{2K}}\,dk \ldots\ldots\ldots\ldots\ldots\ldots\ldots\ldots\ldots\ldots(55).$$

Figure 3. The later definitive formulation elaborating the Boltzmann derivation using the Euler formula for the exponential as the limit of a succession (Equation (51)), leading to continuous Equation (55). The equations are boxed in the Figure.

In the 7th Chapter of his 1938 treatise "Kinetic Theory of Gases", Kennard [31] develops a comprehensive assessment of macroscopic irregular motion of molecules (including, e.g., the Brownian motion) as connected to averaged microscopic fluctuations: the connection between discrete statistical distributions and exponential functions is obtained by the Euler's succession, taking the limit to infinity of the number of particles. Earlier, in a collection of his investigations on statistical mechanics collected in a 1927 book, Tolman [32] had briefly discussed the role of taking limits in the description of fluctuations for a large number of molecules; in his treatise in 1938 [33] the theme of fluctuations and thermodynamic equilibrium are discussed in more details through a detour involving the Stirling formula for factorials and maximization of entropy in the Boltzmann–Gibbs approach. In either way these treatments involved imposing limiting values to specific variables and anticipating the operation that we will discuss in the next section, namely that of taking the thermodynamic limit, see [31,34]: in some cases, as intermediate steps in the course of derivations, physically insightful expressions were encountered.

2.2. The Thermodynamic Limit: The Contribution of Fowler and Collaborators

To give a general foundation to statistical mechanics, stepping stones can be schematized as follows. Darwin and Fowler [10,35] developed their approach in the early twenties, introducing the concept of temperature as the zero principle and defining as key quantities specific distributions and in particular partition functions [10,36]. In a lucid lecture, in 1948 Schrödinger [37] describes their achievements as major after those of Boltzmann (1868) [25] and Gibbs (1902) [12]. A few years later a mathematical analysis around the concept of the thermodynamic limit was carried out by Yang and Lee [38–40]: they considered the limit as to be taken with respect to properties in the neighborhood of phase transitions, and gave a deep theory of associated analytic singularities.

Basic to a variety of modern treatments, the thermodynamic limit concept was mentioned as central in many ways: there are at least three recent books [41–43] and a dedicated paper on statistical mechanics [44] where the treatments are from the very beginning based on the introduction of the concept of thermodynamic limit. Reference [44] refers to the Darwin–Fowler method as powerful alternative to the Boltzmann–Gibbs celebrated construction and describes the Bogoliubov contribution. Technically, Darwin and Fowler [8,9] and Fowler and Guggenheim [35] obtain average quantities from multivariable distributions using an asymptotic method, that is of the steepest descent: reference [45] on p. 53 shows equivalence with taking the Stirling limit for factorials and the Lagrange maximization of functions by undetermined multipliers, a procedure which is standard in the Boltzmann–Gibbs statistical approach to entropy.

Usually, in most popular books the thermodynamic limit is defined only in words. The quantitative definition [46,47] is provided considering extensive variables, the volume V and the number of particles N going to infinity while their ratio, the density $\rho = N/V$, remains finite: see Figure 2 for the reproduction of the original treatment by Boltzmann. The formulas exploit essentially the limit of a succession due to Euler to obtain the exponential function [2]; in books by both Pathria [42] and Huang [40] on statistical mechanics, the very first concept introduced is the thermodynamic limit and provide accessible qualitative, useful presentations of the contributions by Yang and Lee [38–40].

In one of the Landau and Lifshitz series of textbooks, there is a treatment now considered as standard [46]. The section arguably written by L.P. Pitaevski, addresses the problem of fluctuations, obtaining the Poisson distribution considering the volume V of gas occupied by a number of particles N. Let v be a small part of the total volume and proceeding with the same ingredients used by Boltzmann in 1868 ([25] and Figure 2) one can show that the probability for a volume v to contain n molecules follows a Jacob Bernoulli's binomial distribution of the type

$$P(n) = \frac{1}{n!}\frac{N!}{(N-n)!}\left(\frac{v}{V}\right)^n\left(1-\frac{v}{V}\right)^{N-n}. \tag{1}$$

Taking the limit

$$N \to \infty \; and \;\; V \to \infty \tag{2}$$

for an average number of particles $\overline{n}$ in volume V, while the density, namely the ratio

$$\rho = \frac{N}{V} = \frac{\overline{n}}{v} \tag{3}$$

remains finite, the passage from the binomial distribution to the Poisson distribution is obtained by the Euler's formula for the exponential function as a limit of a succession (see also the references [1,41]):

$$P(n) = \frac{\overline{n}^n}{n!} e^{-\overline{n}} \tag{4}$$

This derivation can be taken as representative of the content of the expression "taking the thermodynamic limit".

Other treatments are worthy of mention. Using a path analogous to Boltzmann's "marginal" probability, Eyring and collaborators [14–16] provide an elementary presentation based on a paper by Condon (1938) [34] on a statistical mechanics derivation of the Boltzmann distribution law. The treatment is interesting for chemical kinetics. Considering the equilibrium of a molecular subsystem within a system composed of s harmonic oscillators [15], it is possible to calculate the probability of the molecular subsystem to acquire an energy ε. Assuming a total energy E of the system, one can identify the number of ways to distribute $m = \frac{E-\varepsilon}{h\nu}$ quanta of energy among the s oscillators of the system ($h\nu$ is the energy per quantum):

$$W(s,m) = \binom{s+m-1}{m} = \frac{(s+m-1)!}{m!(s-1)!}. \tag{5}$$

The probability of the molecular system to acquire the energy ε is

$$P(s,m) = \frac{\binom{s+m-1}{m}}{\sum_j \binom{s+j-1}{j}}. \tag{6}$$

Taking what is now recognized as the thermodynamic limit (in this case, s and m tending to infinity) and using the Euler's formula for the exponential function as a limit of a succession, the Boltzmann law is recovered

$$P(\varepsilon_i) = \frac{e^{-\frac{\varepsilon_i}{k_B T}}}{\sum_j e^{-\frac{\varepsilon_j}{k_B T}}}. \tag{7}$$

where ε is energy, k_B is Boltzmann's constant, and T the absolute temperature. In reference [1] modifications needed to obtain Bose–Einstein and Fermi–Dirac distributions are sketched.

As we have seen, it went unnoticed that many formulations had been anticipated by Boltzmann in his 1868 article [25]. Indeed, he himself in the famous paper published in 1877 [1] changed focus, and developed the celebrated procedure: that of searching for most probable values with limits on particle numbers at a given total energy to obtain the entropy by the Lagrange method of undetermined multipliers. Due to this spectacular result the attention of most subsequent investigations was diverted away from the kinetic approach towards thermodynamic treatments. However, in 1940 Hinshelwood [48] sketched a pedagogical justification of Boltzmann's exponential distribution considering the probability of favorable collisions that can lead to a specific energy accumulation into molecules: he relies explicitly on the Euler's limit to obtain the exponential function. See more details in [2], where it is emphasized that when interpreting this mechanism as that operating in the activation of molecules one has a profound insight on chemical reactivity and on rate processes.

2.3. Avoiding the Thermodynamic Limit Describes Nonlinearities of Arrhenius Plots

In chemical kinetics the difficulty occurs in finding macroscopical and canonical kinetics properties, ingredients analogous to those of thermodynamics. Temperature can be introduced assuming the zero principle for the preparation of reactants: the question arises how to use an analogue of the thermodynamic limit when it is arbitrary to define extensive quantities like number of particles (N) or volume (V) in a reactive process.

For kinetics of the chemical and physical rate processes the evidence [49–56] of deviations from Arrhenius behavior is increasing—arguably it is to be associated to moving away significantly from the Boltzmann–Gibbs distribution, particularly going down to low temperatures. In Boltzmann's equation of Figure 2 substituting k/x with $\varepsilon^{\ddagger}\beta/N$ we performed the same mathematical operation as for the thermodynamic limit: the passage of the distribution when n goes to infinity can be written:

$$P_N\left(\varepsilon^{\ddagger}\beta\right) = \left(1 - \frac{\varepsilon^{\ddagger}\beta}{N}\right)^N \overset{N\to\infty}{\to} P\left(\varepsilon^{\ddagger}\beta\right) = e^{-\varepsilon^{\ddagger}\beta} \tag{8}$$

where $\varepsilon^{\ddagger}$ (the activation energy) represents an energetic obstacle for the process to occur and $\beta = 1/k_BT$ is the usual Lagrange multiplier, the "coldness" [2,57]. The exponential Boltzmann–Gibbs distribution $P(\varepsilon^{\ddagger}\beta)$ emerges as a limiting case of a power law distribution $P_N(\varepsilon^{\ddagger}\beta)$ we have shown [1,58–60] that the latter, corresponding to avoiding taking the limit permits that the low temperature deviations in kinetic processes can be described with remarkable consistency in a generality of contexts. This treatment makes explicit the connection of $P_N(\varepsilon^{\ddagger}\beta)$ distribution with Tsallis statistics [61,62] identifying $1/N$ with $1-q$, where N is allowed to be continuous and Tsallis q is classically limited in a small range.

The final expression in (8) is Arrhenius law apart from the pre-exponential factor A: Further convenient introduction of the deformation parameter d in place of $1/N$ one has

$$k_d\left(\varepsilon^{\ddagger}\beta\right) = A\left(1 - d\varepsilon^{\ddagger}\beta\right)^{1/d} \overset{d\to 0}{\to} k\left(\varepsilon^{\ddagger}\beta\right) = e^{-\varepsilon^{\ddagger}\beta} \tag{9}$$

The left-hand side of the correspondence (9) is known as the Aquilanti–Mundim deformed Arrhenius formula. We amply proved that it could be considered uniformly both for $d < 0$ (quantum propensity) and for $d > 0$ (classical propensity). The first case has been treated amply elsewhere [59,63,64]; we mostly focused here on the second case [60,65]. Rarer cases are found for which $d > 0$ and $\varepsilon^{\ddagger} < 0$, and are referred as corresponding to an *anti*-Arrhenius behavior [66–68].

There are some examples in the literature [29,69–71] where there is evidence that our language the classical thermodynamic limit is reached due to the magnitude of Avogadro number $\mathcal{O}(10^{24})$. Considering this number, the statistical mechanics treatment indicates that fluctuations of energy in a canonical ensemble turn out extremely sharp and narrow. This peculiarity of the energy distribution is required to admit the equivalence between averages and most probable values of variables: applicability of statistical techniques to the foundations of thermodynamics relies on this kind of argument, upon which is also based the concept of the thermodynamic limit. At low temperatures, especially in reactive and non-reactive processes, conditions can be violated since the order of magnitude of the activated events is controlled by the Eyring pre-exponential factor $\frac{2\pi}{\hbar\beta} \sim \mathcal{O}(10^{13})$ with a consequent possibility of relaxation of the thermodynamic limit.

Far away from the thermodynamic limit at extreme conditions, a lower thermal limit can be assumed for this *super*-Arrhenius case where chemical or physical rate processes would require an infinite time to proceed, i.e., when

$$1 - d\varepsilon^{\ddagger}\beta \approx 0 \tag{10}$$

this condition provides an interpretation for the parameter *d* already pointed out and defined by us in previous papers [1–3] (and recently confirmed [72,73]):

$$d = \frac{k_B T^{\dagger}}{\varepsilon^{\ddagger}} = \frac{\varepsilon^{\dagger}}{\varepsilon^{\ddagger}} \tag{11}$$

where the superscript † denotes a minimum temperature $T^{\dagger}$ or a thermal energy threshold $\varepsilon^{\dagger}$ for which the kinetic process is operative. The *d* parameter differs from zero as a scale to measure the "thermal limit of propensity" with reference to the lower and higher kinetic energy values respectively to $\varepsilon^{\dagger}$ and $\varepsilon^{\ddagger}$, see Scheme 1: (i) when $\beta \to 0$, and also remaining finite $d = \frac{\varepsilon^{\dagger}}{\varepsilon^{\ddagger}}$ (Cauchy's limit [73]) Equation (9) also tends to the exponential law; and ii) when $\beta \to \infty$, $d\varepsilon^{\ddagger}\beta \to 1$ and $d = \frac{\varepsilon^{\dagger}}{\varepsilon^{\ddagger}}$ remaining finite, a minimum limit can be identified for which the kinetic process may occur. Case ii) applies for $d > 0$, the *super*-Arrhenius cases: the *sub* limit is consistent with Wigner's threshold law for thermoneutral reactions [74–76]. Furthermore, taking advantage of an alternative expansion [77,78] for the Aquilanti–Mundim law

$$k(\beta) = \left(1 - d\varepsilon^{\ddagger}\beta\right)^{\frac{1}{d}} = Ae^{-\varepsilon^{\ddagger}\beta}\left[1 - \tfrac{1}{2}d\left(\varepsilon^{\ddagger}\beta\right)^2 - \tfrac{1}{3}d^2\left(\varepsilon^{\ddagger}\beta\right)^3 - \tfrac{1}{8}(2d-1)d^2\left(\varepsilon^{\ddagger}\beta\right)^4 + \mathcal{O}\left(\beta^5\right)\right] \tag{12}$$

Thermodynamic limit

$\boldsymbol{V \to \infty; N \to \infty;}$ $\quad \boldsymbol{N/V =}$ ***finite*** $\qquad \lim_{\forall\beta;\, d\to 0} A\left(1 - d\varepsilon^{\ddagger}\beta\right)^{1/d} = A\,e^{-\varepsilon^{\ddagger}\beta}$

Kinetic limits

$i)\ \ \beta \to \mathbf{0};\ d = \dfrac{\varepsilon^{\dagger}}{\varepsilon^{\ddagger}} =$ ***finite*** $\qquad \lim_{\forall d;\, \beta\to 0} A\left(1 - d\varepsilon^{\ddagger}\beta\right)^{1/d} = \quad A\,e^{-\varepsilon^{\ddagger}\beta}$

$ii)\ \ \beta \to \infty;\ d\varepsilon^{\ddagger}\beta \to \mathbf{1} \qquad \lim_{\forall d;\, d\varepsilon^{\ddagger}\beta\to 1} A\left(1 - d\varepsilon^{\ddagger}\beta\right)^{1/d} = \begin{cases} 0 \quad (super) \\ A\,\beta^{1/d} \quad (sub) \\ \text{for thermoneutral reactions} \end{cases}$

Scheme 1. The three limits: the thermodynamic limit is an application of that of Euler for $d \to 0$; the kinetic limits apply for extreme values of β confining the temperature range where the process under consideration is operative. For various cases of the *sub*-Arrhenius kinetic limits for $\beta \to \infty$ see text.

Note a misprint in Equation (17) in reference [3]. The three limits previously described are illustrated as defining the thermodynamic and kinetic limits, summarized in the following scheme (see also Figure 4).

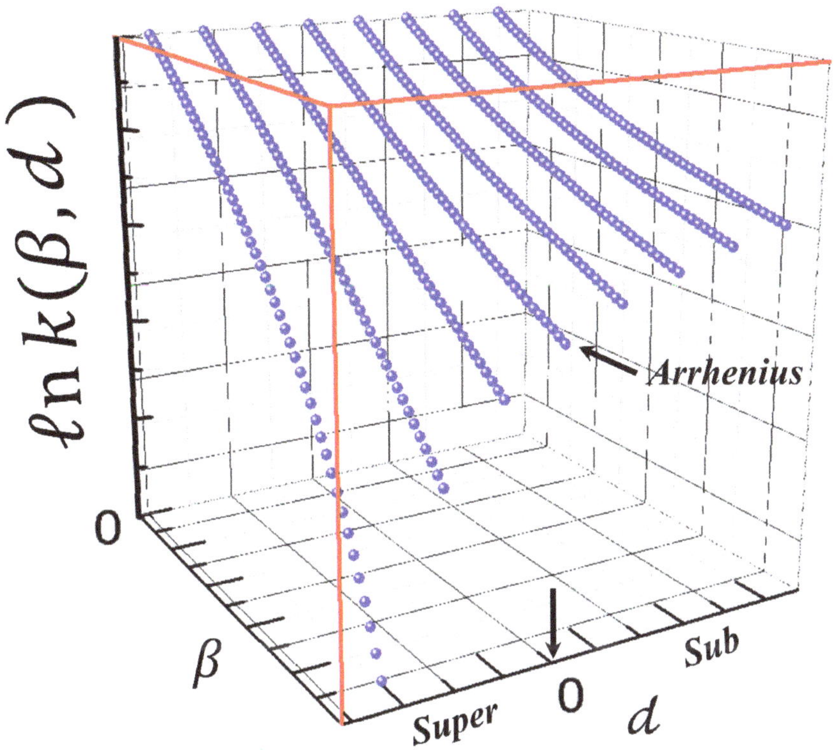

Figure 4. 3D Arrhenius plot illustrating the kinetic limits (Scheme 1) through the functional dependence of the AM rate coefficient k on the parameters d and β. The two kinetic limits apply at the extremes of the range of β. For any value of β, the Arrhenius law is recovered when d tends to zero according to the thermodynamic limit (Scheme 1).

From Figure 4, it is possible to perceive graphically the three limits in a 3D Arrhenius plot, ln$k$ *vs.* d and β. When d tends to zero the Arrhenius law is recovered. For $d > 0$, a convex curvature is generated (*super*-Arrhenius) and the tendency for a lower thermal limit is observed. When $d < 0$ the plot becomes concave (*sub*-Arrhenius) because of quantum mechanically tunneling.

For small tunneling, we showed that [63]

$$d = \frac{-1}{3}\left(\frac{h\nu^{\ddagger}}{2\varepsilon^{\ddagger}}\right)^{2} \tag{13}$$

where $\varepsilon^{\ddagger}$ is the barrier height, directly proportional to $\nu^{\ddagger}$, the square of the frequency for crossing the barrier at the maximum in the potential energy surface. For the concave case, the tendency is attenuated and known as the Wigner limit [74,75] for thermoneutral chemical reactions (see in Scheme 1):

$$\lim_{\forall d;\ d\varepsilon^{\ddagger}\beta\to 1} A\left(1 - d\varepsilon^{\ddagger}\beta\right)^{\frac{1}{d}} = A\,\beta^{\frac{1}{d}}. \tag{14}$$

2.4. *Architecting the Transitivity Concept*

From a conceptual viewpoint and with reference to Figure 1, in a previous paper [2] we emphasized how essentially a statistical mechanics path to chemical kinetics (the theory of change) can be based on a theory of chance, where however criteria for choices need be provided [2,3]: the transitivity concept is exhibited as playing a crucial role.

A recent paper [3] also gives an account of how useful it is the introduction of representations of experimentally or numerically exact rate constant data; the first, of course, is the Arrhenius plane [5], whereby the apparent activation energy is interpreted according to the Tolman's theorem [79]; the second one is the transitivity plane. We sketched here and elaborated elsewhere [4] a further aspect justifying how the definition of the transitivity function that can provide an understanding of microscopic kinetic processes using alternative forms of scaling of the rate data, yielding naturally the conventional statistics used in rate process, from Maxwell–Boltzmann [34] to Tsallis statistics [62], and to the further popular Vogel–Fulcher–Tammann [80–82] distributions.

2.4.1. Tolman's Theorem and the Apparent Activation Energy

Conventionally, starting points in the statistical thermodynamics proposed by Willard Gibbs [12] and Fowler et al. [10,35] are the average energy E of a canonical system obtained as the logarithmic derivative of the partition function Z with respect to β

$$\overline{E} = -\frac{\mathrm{d}}{\mathrm{d}\,\beta} \ln Z \tag{15}$$

When we turn to kinetics, the correspondence can be established considering the average energy to be accumulated by colliding molecules to proceed to reaction. Following Tolman (the first paper is one hundred years old [79]), well-characterized is the concept of apparent activation energy $E_\alpha(\beta)$. This entity is customarily obtained by chemical kinetics data on reaction rate coefficients $k(\beta)$ $k(\beta)$, phenomenologically from the Arrhenius plot (as recommended in 1996 by the definition of the International Union of Pure and Applied Chemistry [83]):

$$E_a(\beta) = -\frac{\mathrm{d}}{\mathrm{d}\,\beta} \ln k(\beta) \tag{16}$$

The apparent activation energy can be written as the difference between the average internal energy of the reacting molecules and that of all molecules in the system: this is the content of statement of the so-called Tolman's theorem [35,79], which has been analyzed quantum mechanically by Fowler and Guggenheim [35]: the meaning is that E_a represents a measure of an energetic obstacle to the progress of the reaction, reinterpreted subsequently and exploited as the barrier height energy in Eyring's formulation of the transition state theory [13]. Basic is to consider the apparent activation energy as essentially the $\varepsilon^{\ddagger}$ parameter of the previous section (actually, the double dagger notation is that introduced by Eyring).

2.4.2. Planck Black-Body Radiation and Reciprocal Energy

We provided now a further perspective viewpoint of the phenomenological parameters involved in the definition of the activation energy and its reciprocal, the transitivity function, going back to the elementary formulation of Planck to solve the problem of the average energy of a black body [14,84]. Assuming a system composed of harmonic oscillators with quantized energy $\epsilon_n = nh\nu$, where ν is the frequency of the oscillator, the partition function for this system is given as,

$$Z = \sum\nolimits_n e^{-\epsilon_n \beta} \tag{17}$$

The total average energy of this system can be calculated using Equation (15)

$$\begin{aligned}
\overline{E} &= -\tfrac{d}{d\,\beta}\ln\textstyle\sum_n e^{-\epsilon_n\beta} \\
\overline{E} &= -\tfrac{d}{d\,\beta}\ln\left[1+e^{-h\nu\beta}+\left(e^{-h\nu\beta}\right)^2+\left(e^{-h\nu\beta}\right)^3+\ldots\right] \\
\overline{E} &= -\tfrac{d}{d\,\beta}\ln\left[\tfrac{1}{1-e^{-h\nu\beta}}\right] = \tfrac{h\nu}{e^{h\nu\beta}-1}
\end{aligned} \tag{18}$$

At low temperatures $(\beta \to \infty)$, expansion of the reciprocal of $\overline{E}$ assumes a functional form that produces a power law in β,

$$\frac{1}{\overline{E}} = \frac{1}{h\nu}\left[1 + h\nu\beta + \frac{1}{2!}(h\nu\beta)^2 + \frac{1}{3!}(h\nu\beta)^3 + \frac{1}{4!}(h\nu\beta)^4 + \ldots\right] \tag{19}$$

suggesting the usefulness of introducing also to this case the inverse of $\overline{E}$, as an analog of the transitivity function $\gamma(\beta)$. See [1] for elaboration of quantum statistical treatments.

2.4.3. Activation and Transitivity: A Prototypical Unimolecular Reaction Model

Another interesting case that generates a functional form justifying the introduction of transitivity is the model considered in the Twelfth Chapter of reference [35] by Fowler and Guggenheim: they propose the initial steps of a kinetic theory for unimolecular processes, starting from a quantum theoretical formulation of Tolman's theorem for calculating the probability for molecules to react after acquiring a sufficient amount of energy $\varepsilon^{\ddagger}$ distributed over the s internal degrees of freedom characterizing the reacting molecule. The model, arguably valid within a small enough neighborhood of $\beta = 0$, is that all active molecules have the same probability of decomposition, requiring an average over all possible values of energy for the active molecules: these hypotheses represent the prototypical theory of unimolecular reactions further elaborate into that of Slater [85] and to the more successful RRKM formulation. We find interesting to elaborate further and provide a continuation of their formulation.

Consider, as first suggested by René Marcelin [86], that a molecule is "active" if the energy exceeding $\varepsilon^{\ddagger}$ is distributed over s internal vibrational degree of freedom: the last formula of Fowler and Guggenheim presentation for the unimolecular rate constant can be transcribed here in our notation in a remarkably simplified form

$$k(\beta) = Ae^{-\varepsilon^{\ddagger}\beta}\sum_{r=0}^{s-1}\frac{1}{r!}\left(\varepsilon^{\ddagger}\beta\right)^r. \tag{20}$$

First, we note that the sum can be given in closed form

$$k(\beta) = A\left(\frac{e^{\varepsilon^{\ddagger}\beta}\Gamma(s,\varepsilon^{\ddagger}\beta)}{\Gamma(s)}\right)e^{-\varepsilon^{\ddagger}\beta}. \tag{21}$$

where $\Gamma(s,\varepsilon^{\ddagger}\beta)$ is the incomplete Γ-function. Additionally, we calculated the activation energy for the Fowler–Guggenheim model by logarithmic differentiation of the rate coefficient (21) with respect to β

$$E_a = -\frac{\partial \ln k}{\partial \beta} = \varepsilon^{\ddagger} - \frac{\sum_{r=1}^{s-1}\frac{1}{(r-1)!\beta}\left(\varepsilon^{\ddagger}\beta\right)^r}{\sum_{r=0}^{s-1}\frac{1}{r!}\left(\varepsilon^{\ddagger}\beta\right)^r} = \varepsilon^{\ddagger}\frac{\left(\varepsilon^{\ddagger}\beta\right)^{s-1}}{\Gamma(s,\varepsilon^{\ddagger}\beta)}e^{-\varepsilon^{\ddagger}\beta}. \tag{22}$$

Finally, we got a closed form in β also for the transitivity function

$$\gamma(\beta) = \frac{1}{E_a} = (s-1)!\sum_{r=0}^{s-1}\frac{1}{r!\beta}\left(\varepsilon^{\ddagger}\beta\right)^{r-s} = \frac{1}{\varepsilon^{\ddagger}}\frac{\Gamma(s,\varepsilon^{\ddagger}\beta)}{\left(\varepsilon^{\ddagger}\beta\right)^{s-1}}e^{\varepsilon^{\ddagger}\beta}. \tag{23}$$

This model for unimolecular reactions is of interest for any pseudo-first-order processes. Two other cases analyzed by reference [35] should be similarly investigated, permitting asymptotic analysis through well-known properties of special functions.

3. Scaling in the Transitivity Plane

3.1. *Transitivity and Renormalization Group Coupling*

It is important to recognize the similarity between the functional form of transitivity γ with respect to the rate coefficient k and the reciprocal temperature β (Figure 1); and the renormalization group coupling parameter, credited to Callan [87] and Symanzik [88] (see also Wilson [89]): $\beta_{CS}(g) = -(\partial g/\partial \ln \mu)$ defines modernly the relationship between the coupling constant g and the energy scaling function μ. The equation encodes the mathematical apparatus in both quantum field theory and the theories of critical phenomena used to handle problems with singularities, such as those occurring at phase transitions. See the lucid presentation by Weinberg [90] (see also [91,92]).

3.2. *Classes of Universal Behaviors*

The previous sections have shown that from a kinetic point view, the reciprocal of the activation energy can be properly defined as the transitivity γ, specific of a process and interpreted as a measure of the propensity for the reaction to proceed. Our notations stem from the fact that the transitivity can take a gamma of values smooth as function of β in a sufficiently ample range of temperatures. Its limiting values will serve to localize any abrupt changes, e.g., in mechanisms of processes or in phase transitions. Generally, if a Laurent expansion defined in references [1–4] is assumed to hold in a neighborhood around a reference value denoted as β_0, it behaves asymptotically as

$$\gamma(\beta) \sim \alpha\,(\beta - \beta_0)^{\zeta} \tag{24}$$

General series for $\gamma(\beta)$ where previously given in reference [1,3]. Now, the transitivity plane, γ *vs.* β, (see Figure 3) can be interpreted as confining the range of existence of a system between limiting temperatures in consonance with the thermal kinetic limits defined in Section 2.2. The two temperatures or limiting coldnesses β are generally contained between the extremes $\beta \approx 0$ to $\beta \approx \beta^{\dagger}$ defining the temperature window where a process is operative. The simplest model for γ is a linear path from $\alpha = 1/\varepsilon^{\ddagger}$ to $\beta^{\dagger} = 1/\varepsilon^{\ddagger}$ according to the AM formula [3]. In fact, the limiting formula derived from Equation (24) in reference [3] yields

$$\gamma(\beta) = \frac{1}{\varepsilon^{\ddagger}}\left(1 - \frac{\beta}{\beta^{\dagger}}\right). \tag{25}$$

It is interesting to express known temperature-dependence rate laws generalizing the previous equation as

$$\gamma(\beta) = \frac{1}{\varepsilon^{\ddagger}}\left(1 - \frac{\beta}{\beta^{\dagger}}\right)^{\zeta} \tag{26}$$

The exponent ζ = 0, 1, and 2 generates the Arrhenius, Aquilanti–Mundim (AM), and Vogel–Fulcher–Tammann (VFT) laws, respectively [3]. Many other paths can serve as models for the transitivity function for different values of ζ (see Figure 5). Generalization to non-integer values shows perspectives of correlation with critical exponents in mode coupling theory and with universality classes of kinetic transitions (see also Section 4). Studies in the glass transition field show [93–95] that systems with a large fragility (strong non-Arrhenius behavior) present ranges of universality separated by a crossover temperature: in some works considering glass-forming systems and, e.g., for the prototypical reaction F + H_2 (D_2) at low and ultra-low temperatures [20], should permit to categorize the universality classes in a wide temperature range by the critical exponent ζ, possibly empirically a non-integer.

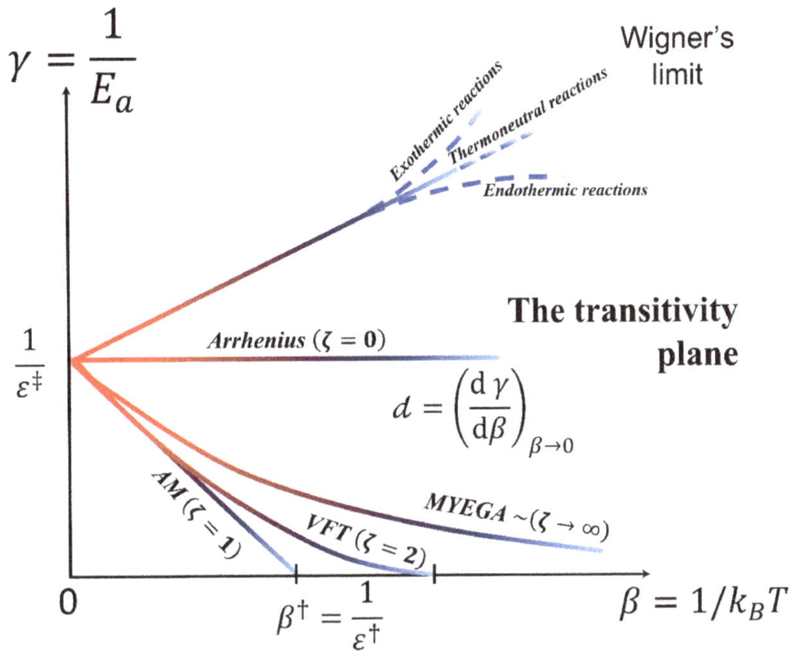

Figure 5. The alpha-zeta totem. The transitivity plane, introduced in [3], $\gamma = 1/E_a$ *vs.* $\beta = 1/k_BT$ serves to give a proper scaling to the phenomenological parameters occurring in the study of nonlinear Arrhenius plots. The Arrhenius behavior is given as corresponding to a line parallel to the β axis starting at $1/\varepsilon^{\ddagger}$. Deviations from the Arrhenius behavior give the transitivity function γ a straight-line at small β, which it is connected to the d parameter of the Aquilanti–Mundim (AM) law. At low temperatures, the transitivity function tends to characteristic ultra-cold limiting values: (i) for $d < 0$ (*sub*-Arrhenius) it tends to the Wigner limit and (ii) for $d > 0$ (*super*-Arrhenius), γ, namely the propensity for reaction to occur, vanishes in $\beta^{\dagger}$ from Mauro–Yue–Ellison–Gupta–Allan (MYEGA) [96], Aquilanti–Mundim (AM) [60] and Vogel–Fulcher–Tammann (VFT) [80–82] path. In [3] $1/\varepsilon^{\ddagger}$ is defined as α, so that the transitivity plane provides a primitive totem for elementary chemical kinetics, employing the first six letters of the Greek alphabet, $\alpha\ \beta\ \gamma\ \delta\ \varepsilon\ \zeta$ for the parameters, not all independent (see text).

In reference [3], we show cases when curvature in the Arrhenius plot can be linearized. Interestingly, a formulation was empirically proposed in 1980s to fit the temperature dependence of properties of glass-forming materials [97–99]. Here, the proposed connection through the Tolman's theorem is assumed as a scaling tool for relaxation processes: the relationships appearing in the transitivity plane turn out as explicitly universal for the linear dependence in β, at least in a significantly wide neighborhood near the origin of $\beta \gtrsim 0$. Most important is that all parameters are given both a physical meaning and the possibility of being estimated by physical models.

4. Perspectives on Rate Processes from the Arrhenius and the Transitivity Planes

There are a variety of chemical reactions and rate processes that deviate from Arrhenius behavior, and this list of them is currently expanding upon consideration of several types of phenomena being documented [1,59,60,100,101]. Below, we collected three important examples in which a systematic investigation of the universality is in progress.

(i) The rates of biological processes are strongly affected at low temperatures by deviations from Arrhenius law; however there are large uncertainties especially when quantifying, as usual these deviations using the "Arrhenius Break Temperature" assumption, see previous discussions [3,60].

The difficulty of identifying a transition temperature in the Arrhenius plot for the respiration processes [60,102,103] can be easily overcome using the transitivity plot, emphasizing sudden transitions described within the Aquilanti–Mundim law (universality class with $\zeta = 1$).

(ii) Further applications concern the glass transition phenomenon occurring in a variety of materials: This is considered one of the most complex open problems in condensed matter physics. In the neighborhood of the glass transition temperature, the kinetic coefficients—diffusion, viscosity, and relaxation time—present deviations from the Arrhenius law specifically depending on the material composition. In reference [4], we examined the nonlinear temperature dependence of the relaxation time of propylene carbonate [98,99] from the transitivity plot: it is presented a perspective tool to observe a transition temperature connecting regimes described by two Aquilanti–Mundim straight lines in transitivity plane, and identifying the crossover temperature [93,94].

(iii) Among phenomena akin to glass transitions but on extremely larger timescales, very important are those occurring in geochemical environments, where nonlinearity of the temperature dependence of the viscosity of rocks is often observed in the Arrhenius plots. In Figure 6, the nonlinearity in Arrhenius plot for **Cl_OF** silicate [49,104,105] also obeys the Aquilanti–Mundim law when analyzed in the transitivity plot: however, no transition temperature is revealed in this case.

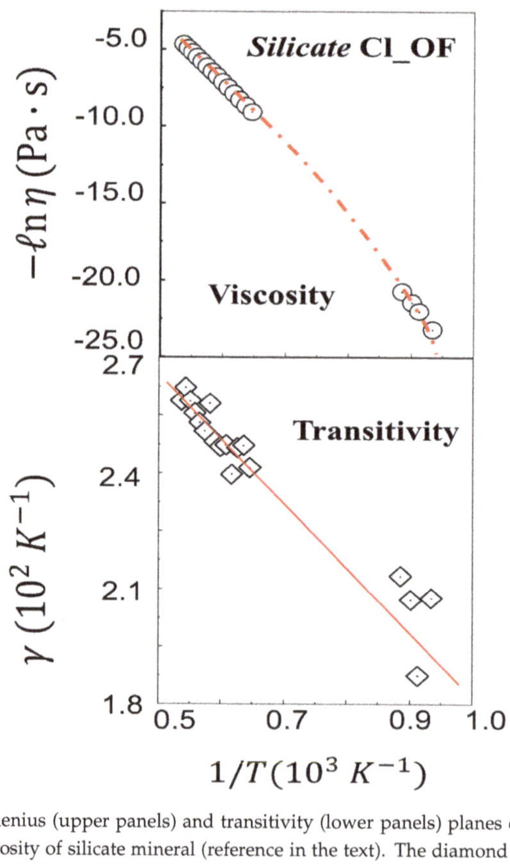

Figure 6. The Arrhenius (upper panels) and transitivity (lower panels) planes of the temperature dependence of viscosity of silicate mineral (reference in the text). The diamond symbols represent the transitivity values obtained numerically. Red lines emphasize region where the temperature dependence of the transitivity is linearized, as expected by the Aquilanti–Mundim law.

In the exemplary processes presented above, the scaling provided by the transitivity variable makes explicit the corresponding extent of deviations from Arrhenius law, emphasizing kinetic transition temperatures when they appear. Computational tools for assisting in assessing this behavior are presented in a code described in a companion article in this topical collection [4].

5. Conclusions and Outlook

The approaches to the description of rate processes formulation have evolved due to the synergism between phenomenological and computational approaches: with Arrhenius law representing the former and the transition-state theory standing at the foundations of the latter. However, with the advance of experimental and computational techniques, these approaches needed extensions able to cope with new problems, such as quantum effects (e.g., tunneling and resonance) in atomic and molecular systems, stochastic motion of particles in condensed environment, non-equilibrium effects in classical and quantum formulations. From several modern techniques for treating kinetic problems, we can cite Feynman-like path integral formulations [106–108] to estimate temperature dependence of rate constants in chemical reactions, mode-coupling theory [109] for describing the physics of glass formation; and the development of rational extended thermodynamics [23,110] to treat systems far away from equilibrium.

The implementation of modern formulations to new experimental data and computational simulations requires a complex set of microscopic information to estimate kinetic parameters, making formidable the problem of describing many-particle dynamics and kinetic equations. Concomitantly, phenomenological approaches continue to be important pillars for the enhancement of ab initio formulations beyond: Arrhenius [5], Vogel–Fulcher–Tammann [80–8582 Williams–Landel–Ferry [111], Power law [112], Bässler [113], and Nakamura–Takayanagi–Sato [114] laws are a sequence of useful models to describe problems in extreme and highly complex environments. Motivated by this perception and establishing connection with Tsallis statistics [62] for classical propensities, in the last ten years we have worked in close synergism between phenomenological and ab initio or semiempirical formulations. A key guide came by Euler's expression for the exponential function as a limit of succession, a formulation accompanied by physicochemical meanings originally suggested for gas kinetic theory and chemical kinetics processes.

The "prequel" to the saga has been reconstructed in Section 2. We recapitulate the steps that originated essentially from following the Maxwell–Boltzmann path and involving at some stage application of the Euler's formula: Boltzmann (1868) [25] (Figure 2) was the first that succeeded to prove the Maxwell's distribution working with marginal probabilities in what is now called the thermodynamic limit; subsequently, Maxwell (1879) [27] (Figure 3) and Jeans (1916) [28] developed rigorous formulations performing mathematically the Euler's formula for the thermodynamic limit; Uhlenbeck and Goudsmit [29] in their study of finiteness of particle number stated clearly that their formulation is written in the spirit of the Maxwell–Boltzmann original treatments. In these formulations at the final stage always the Euler's limit is invoked, by which the exponential distribution function is recovered upon taking the thermodynamic limit.

Connection with concomitant modern approaches is relevant. Recently, it has been asserted that the molecular world and its reactivity can be interpreted by theories involving Fuzzy sets and Fuzzy logic [115]. These theories have been formulated by the electronic engineer Lotfi Zadeh, and are useful to model how humans "compute" by using words [116]. Every word of the natural language, represented by a Fuzzy set, is like a "quantum" of information, whose meaning is context-dependent. Similarly, every molecule or every atom of the microscopic world is like a Fuzzy set, i.e., like a word of the "molecular language". Every molecule can exist as a collection of different conformers and every atom as a superposition of different quantum states. These "molecular Fuzzy sets" show context-dependent behaviors.

The connection with modern statistical mechanics appears to emerge as follows. In our context, it is needed to understand the meaning of the deviations from the Arrhenius equation. Let us cite from the incipit of [41]:

In statistical mechanics we are concerned with the physical properties of large systems. We assume the existence of the thermodynamic limit (a main concern in this paper, Section 2). The peculiarity, which requires that the mechanics of such a system is "statistical", stems from the fact that such a system is as a rule incompletely defined. By this we mean that the equations of motion for such a system cannot be uniquely solved. Were this true in Gibbs' time already for the simple reason of mathematical complexity, the real problem is not computational, as is clear from interesting computer simulations currently available. Basically, the need for statistical methods stems from the lack of detailed information on the system.

The chain of emerging connections continues. The interpretation of the experimental evidence might require Zener's geometric programming optimizations [117]: geometric programming is a nonlinear mathematical optimization method used to minimize functions that are in the form of polynomials subject to constraints of the same type. The connection between geometric programming and the Darwin–Fowler method has been established since some time [117] (see also a modern approach [118]). Since the data used in the optimization procedure are always affected by errors and uncertainties, a strategy to handle them is provided by the theory of Fuzzy sets, as discussed very recently [119], for example in reference [4], in generally in most of our work we used the generalized simulated annealing (GSA) [120]. The application of Fuzzy optimization algorithms can avoid rigidity and stiffness and reduce information loss arising from the conventional optimization procedures of statistical mechanics.

Author Contributions: Conceptualization, V.H.C.-S., N.D.C. and V.A.; Investigation, V.H.C.-S.; Methodology, V.H.C.-S., N.D.C. and V.A.; Writing—original draft, V.H.C.-S., N.D.C. and V.A.; Writing—review & editing, V.H.C.-S., N.D.C. and V.A. All authors have read and agreed to the published version of the manuscript.

Funding: Valter H. Carvalho-Silva thanks Brazilian agency CNPq for the research funding programs (Universal 01/2016—Faixa A—406063/2016-8) and Organizzazione Internazionale Italo-latino Americana (IILA) for a Biotechnology Sector-2019 scholarship. Nayara D. Coutinho and Vincenzo Aquilanti acknowledge the Italian Ministry for Education, University and Research, MIUR, for financial support: SIR 2014 "Scientific Independence for young Researchers" (RBSI14U3VF) coordinated by Dr. Federico Palazzetti.

Conflicts of Interest: The authors declare that the research was conducted in the absence of any commercial or financial relationships that could be construed as a potential conflict of interest.

Appendix A

Deformation of Statistical Distributions of Molecular Velocities and Kinetic Energies

The previous discussion of the so-called Tolman's theorem [79] provides statistical mechanics meaning to the energetics of an activation process: here we present a derivation of the fraction of particles that exceed a certain amount of energy for a distribution of velocities far from the thermodynamic limit and deforming the Maxwell distribution. To calculate the probability distribution of a particle to acquire a certain velocity described by the vector v we used the formulation presented in references [121,122].

$$P(v) = \mathcal{N}_d \left(\frac{m\beta}{2\pi}\right)^{\frac{3}{2}} \left(1 - d\frac{m\beta}{2}v^2\right)^{\frac{1}{d}}, \tag{A1}$$

where m is the reactant mass, d can be interpreted as the parameter defined in the Section 2.2. and normalization is given by

$$\mathcal{N}_d = \left\{ \begin{array}{c} (-d)^{\frac{3}{2}} \frac{\Gamma(-\frac{1}{d})}{\Gamma(-\frac{1}{d}-\frac{3}{2})}, \; for\; d < 0 \\ \frac{1}{4} d^{\frac{1}{2}} (2+d)(2+3d) \frac{\Gamma(\frac{1}{2}+\frac{1}{d})}{\Gamma(\frac{1}{d})}, \; for\; d > 0 \end{array} \right\}, \tag{A2}$$

when $\mathcal{d} \to 0$, $\mathcal{N}_{\mathcal{d}} \to 1$ and the Maxwell–Boltzmann molecular speed distribution is recovered [121,122].

To account for the alternatives ways to combine the velocity vectors, one can follow Maxwell obtaining the variety of states available to be accessed by performing an integral over the surface of the area of a sphere of radius $v = |v|$, $4\pi v^2$. From Equation (A2), the distribution assumes the following form,

$$P(v)dv = 4\pi\mathcal{N}_{\mathcal{d}}\left(\frac{m\beta}{2\pi}\right)^{\frac{3}{2}} v^2\left(1 - \mathcal{d}\frac{m\beta}{2}v^2\right)^{\frac{1}{\mathcal{d}}} \mathrm{d}v, \tag{A3}$$

which can also be written as a function of the available energy (E) to be accumulated in the system [123],

$$P(E)dE = \frac{2\beta^{\frac{3}{2}}}{\pi^{\frac{1}{2}}}\mathcal{N}_{\mathcal{d}}E^{\frac{1}{2}}(1 - \mathcal{d}E\beta)^{\frac{1}{\mathcal{d}}}\,\mathrm{d}E. \tag{A4}$$

In order to calculate the fraction of molecules, $\mathrm{d}N/N$, which accumulates energy between E and $E + \mathrm{d}E$ during the collisional processes, we can use Equation (A4),

$$\frac{dN}{N} = \frac{2\beta^{\frac{3}{2}}}{\pi^{\frac{1}{2}}}\mathcal{N}_{\mathcal{d}}E^{\frac{1}{2}}(1 - \mathcal{d}E\beta)^{\frac{1}{\mathcal{d}}}\,\mathrm{d}E. \tag{A5}$$

The fraction of molecules $F^{\ddagger}$ with energy in excess energy of a certain specified value $\varepsilon^{\ddagger}$ is relevant for chemical kinetics problems and can be expressed in closed form calculating the following integral

$$F^{\ddagger} = \frac{2\beta^{\frac{3}{2}}}{\pi^{\frac{1}{2}}}\mathcal{N}_{\mathcal{d}}\int_{\varepsilon^{\ddagger}}^{\infty} E^{\frac{1}{2}}(1 - \mathcal{d}E\beta)^{\frac{1}{\mathcal{d}}}\mathrm{d}E = \frac{2\beta^{\frac{3}{2}}}{\pi^{\frac{1}{2}}}\mathcal{N}_{\mathcal{d}}\varepsilon^{\ddagger\,\frac{3}{2}}\mathcal{F}\left(\frac{3}{2}, -\frac{1}{\mathcal{d}}; \frac{5}{2}; \mathcal{d}\varepsilon^{\ddagger}\beta\right). \tag{A6}$$

where $\mathcal{F}$ is the hypergeometric function of Gaussian exemplary numerical simulations of the transitivity function that can be calculated from Equation (A6) and will be presented elsewhere. Being in closed form and containing functionality well characterized mathematically, Equation (A6) should be useful for asymptotic analysis of the physical features of the "deformed" distribution.

References

1. Aquilanti, V.; Coutinho, N.D.; Carvalho-Silva, V.H. Kinetics of Low-Temperature Transitions and Reaction Rate Theory from Non-Equilibrium Distributions. *Philos. Trans. R. Soc. London A* **2017**, *375*, 20160204. [CrossRef] [PubMed]
2. Aquilanti, V.; Borges, E.P.; Coutinho, N.D.; Mundim, K.C.; Carvalho-Silva, V.H. From statistical thermodynamics to molecular kinetics: the change, the chance and the choice. *Rend. Fis. Acc. Lincei.* **2018**, *28*, 787–802. [CrossRef]
3. Carvalho-Silva, V.H.; Coutinho, N.D.; Aquilanti, V. Temperature dependence of rate processes beyond Arrhenius and Eyring: Activation and Transitivity. *Front. Chem.* **2019**, *7*, 380. [CrossRef]
4. Machado, H.G.; Sanches-Neto, F.O.; Coutinho, N.D.; Mundim, K.C.; Palazzetti, F.; Carvalho-Silva, V.H. "Transitivity": a code for computing kinetic and related parameters in chemical transformations and transport phenomena. *Molecules* **2019**, *24*, 3478. [CrossRef]
5. Arrhenius, S. On the reaction rate of the inversion of non-refined sugar upon souring. *Z. Phys. Chem.* **1889**, *4*, 226–248.
6. van't Hoff, J.H. *Études de Dynamique Chimique*; Frederik Muller: Amsterdam, The Netherlands, 1884.
7. Laidler, K.J. The development of the Arrhenius equation. *J. Chem. Educ.* **1984**, *61*, 494. [CrossRef]
8. Darwin, C.G.; Fowler, R.H. XLIV. On the partition of energy. *Philos. Mag. J. Sci.* **1922**, *44*, 450–479. [CrossRef]
9. Darwin, C.G.; Fowler, R.H. LXXI. On the partition of energy—Part II. Statistical principles and thermodynamics. *Philos. Mag. J. Sci.* **1922**, *44*, 823–842. [CrossRef]
10. Fowler, R.H. *Statistical Mechanics: The Theory of the Properties of Matter in Equilibrium*; The University Press: Cambridge, UK, 1929.

11. Brush, S.G.; Hall, N.S. The Kinetic Theory of Gases: An Anthology of Classic Papers with Historical Commentary. In *History of Modern Physical Sciences*; Imperial College Press: London, UK, 2003; Volume 1.
12. Gibbs, J.W. *Elementary Principles in Statistical Mechanics: Developed with Especial Reference to the Rational Foundation of Thermodynamics*; C. Scribner's Sons: New York, NY, USA, 1902.
13. Glasstone, S.; Laidler, K.J.; Eyring, H. The Theory of Rate Processes: The Kinetics of Chemical Reactions, Viscosity, Diffusion and Electrochemical Phenomena. In *International Chemical Series*; McGraw-Hill: New York, NY, USA, 1941.
14. Eyring, H.; Walter, J. An elementary formulation of statistical mechanics. *J. Chem. Educ.* **1941**, *18*, 73. [CrossRef]
15. Walter, J.E.; Eyring, H.; Kimball, G.E. *Quantum Chemistry*; Wiley & Sons: New York, NY, USA, 1944.
16. Johnson, F.H.; Eyring, H.; Polissar, M.J. *The Kinetic Basis of Molecular Biology*; Wiley & Sons: New York, NY, USA, 1954.
17. Maxwell, J.C. XX. Illustrations of the dynamical theory of gases. *Philos. Mag.* **1860**, *19*, 377–391. [CrossRef]
18. Maxwell, J.C. On the Dynamical Theory of Gases. *Philos. Trans.* **1866**, *157*, 49–88.
19. Aquilanti, V.; Cavalli, S.; Fazio, D. De; Volpi, A.; Aguilar, A.; Lucas, J.M. Benchmark rate constants by the hyperquantization algorithm. The F + H_2 reaction for various potential energy surfaces: features of the entrance channel and of the transition state, and low temperatur e reactivity. *Chem. Phys.* **2005**, *308*, 237–253. [CrossRef]
20. De Fazio, D.; Aquilanti, V.; Cavalli, S. Quantum Dynamics and Kinetics of the F + H_2 and F + D_2 Reactions at Low and Ultra-Low Temperatures. *Front. Chem.* **2019**, *7*, 328. [CrossRef]
21. De Fazio, D. The H + $HeH^+ \rightarrow$ He + H_2^+ reaction from the ultra-cold regime to the three-body breakup: Exact quantum mechanical integral cross sections and rate constants. *Phys. Chem. Chem. Phys.* **2014**, *16*, 11662–11672. [CrossRef]
22. Cavalli, S.; De Fazio, D.; Aquilanti, V. Quantum effects in the F+H_2 and F+D_2. *Front. Chem.* **2019**, *7*, 328.
23. Ruggeri, T.; Sugiyama, M. Rational extended thermodynamics: a link between kinetic theory and continuum theory. *Rend. Fis. Acc. Lincei* **2020**, *31*, 33–38. [CrossRef]
24. Cercignani, C. *Ludwig Boltzmann: The Man Who Trusted Atoms*; Oxford University Press: Oxford, UK, 2006.
25. Boltzmann, L. Studien über das Gleichgewicht der lebendigen Kraft zwischen bewegten materiellen Punkten. *Wien. Ber.* **1868**, *58*, 517–560.
26. Costantini, D.; Garibaldi, U. A Probabilistic Foundation of Elementary Particle Statistics. Part I. *Stud. Hist. Philos. Sci. B Stud. Hist. Philos. Mod. Phys.* **1998**, *28*, 483–506. [CrossRef]
27. Maxwell, J.C. On Boltzmann's theorem on the average distribution of energy in a system of material points. *Trans. Cambridge Phil. Soc.* **1879**, *12*, 713–741.
28. Jeans, J. *The Dynamical Theory of Gases*; Cambridge University Press: Cambridge, UK, 1916.
29. Uhlenbeck, G.E.; Goudsmit, S. Statistical energy distributions for a small number of particles. *Zeeman Verhandenlingen* **1935**, *Martinus N*, 201–211.
30. Adib, A.B.; Moreira, A.A.; Andrade Jr, J.S.; Almeida, M.P. Tsallis thermostatistics for finite systems: a Hamiltonian approach. *Phys. A Stat. Mech. its Appl.* **2003**, *322*, 276–284. [CrossRef]
31. Kennard, E.H. Kinetic Theory of Gases: With an Introduction to Statistical Mechanics. In *International Series in Pure and Applied Physics*; McGraw-Hill: New York, NY, USA, 1938.
32. Tolman, R.C. Statistical Mechanics with Applications to Physics and Chemistry. *Chem. Eng. News* **1927**, *5*, 16.
33. Tolman, R.C. *The Principles of Statistical Mechanics*; Oxford University: London, UK, 1938.
34. Condon, E.U. A Simple Derivation of the Maxwell-Boltzmann Law. *Phys. Rev.* **1938**, *54*, 937–940. [CrossRef]
35. Fowler, R.H.; Guggenheim, E.A. *Statistical Thermodynamics: A Version of Statistical Mechanics for Students of Physics and Chemistry*; Macmillan: London, UK, 1939.
36. Fowler, R.; Guggenheim, E.A. *Statistical Thermodynamics*; Cambridge University Press: Cambridge, UK, 1939.
37. Schrodinger, E. *Statistical Thermodynamics: A Course of Seminar Lectures*; Cambridge University Press: London, UK, 1948.
38. Yang, C.N.; Lee, T.D. Statistical theory of equations of state and phase transitions. I. Theory of condensation. *Phys. Rev.* **1952**. [CrossRef]
39. Lee, T.D.; Yang, C.N. Statistical theory of equations of state and phase transitions. II. Lattice gas and ising model. *Phys. Rev.* **1952**. [CrossRef]
40. Huang, K. *Statistical Mechanics*, 2nd ed.; John Wiley and Sons: New York, NY, USA, 1987.

41. Van Vliet, C.M. *Equilibrium and Non-Equilibrium Statistical Mechanics*; World Scientific: Singapore, 2008.
42. Pathria, R.K.; Beale, P.D. *Statistical Mechanics*, 3rd ed.; Academic Press: Cambridge, MA, USA, 2007.
43. Blundell, S.J.; Blundell, K.M. *Concepts in Thermal Physics*; Oxford University Press: Oxford, UK, 2010.
44. Kuzemsky, A.L. Thermodynamic Limit in Statistical Physics. *Int. J. Mod. Phys. B* **2014**, *28*, 1430004. [CrossRef]
45. Hill, T.L. *Statistical Mechanics: Principles and Selected Applications*; McGraw-Hill: New York, NY, USA, 1956.
46. Landau, L.D.; Lifshitz, E.M. *Statistical Physics*, 5th ed.; Pergamon Press: Oxford, UK, 1958.
47. Lavenda, B.H. *Statistical Physics: A Probabilistic Approach*; Dover Publications: Mineola, NY, USA, 2016.
48. Hinshelwood, C.N. The kinetics of chemical change. *J. Chem. Educ.* **1940**, *39*, 79.
49. Giordano, D.; Russell, J.K.; Dingwell, D.B. Viscosity of magmatic liquids: A model. *Earth Planet. Sci. Lett.* **2008**, *271*, 123–124. [CrossRef]
50. Angell, C.A. Formation of glasses from liquids and biopolymers. *Science* **1995**, *267*, 1924–1935. [CrossRef]
51. Kohen, A.; Cannio, R.; Bartolucci, S.; Klinman, J.P. Enzyme dynamics and hydrogen tunnelling in a thermophilic alcohol dehydrogenase. *Nature* **1999**, *399*, 496–499. [CrossRef]
52. Limbach, H.-H.; Miguel Lopez, J.; Kohen, A. Arrhenius curves of hydrogen transfers: tunnel effects, isotope effects and effects of pre-equilibria. *Philos. Trans. R. Soc. Lond. B. Biol. Sci.* **2006**, *361*, 1399–1415. [CrossRef] [PubMed]
53. Smith, I.W.M. The temperature-dependence of elementary reaction rates: beyond Arrhenius. *Chem. Soc. Rev.* **2008**, *37*, 812–826. [CrossRef]
54. Capitelli, M.; Pietanza, L.D. Past and present aspects of Italian plasma chemistry. *Rend. Fis. Acc. Lincei* **2019**, *30*, 31–48. [CrossRef]
55. Kubo, K. A view on the break in the Arrhenius plots. *J. Theor. Biol.* **1985**, *115*, 551–569. [CrossRef]
56. Nishiyama, M.; Kleijn, S.; Aquilanti, V.; Kasai, T. Temperature dependence of respiration rates of leaves, 18O-experiments and super-Arrhenius kinetics. *Chem. Phys. Lett.* **2009**, *482*, 325–329. [CrossRef]
57. Müller, I. The coldness, a universal function in thermoelastic bodies. *Arch. Ration. Mech. Anal.* **1971**, *41*, 319–332. [CrossRef]
58. Vieira, C.L.; Sanches Neto, F.O.; Carvalho-Silva, V.H.; Signini, R. Design of apolar chitosan-type adsorbent for removal of Cu(II) and Pb(II): An experimental and DFT viewpoint of the complexation process. *J. Environ. Chem. Eng.* **2019**, *7*, 103070. [CrossRef]
59. Silva, V.H.C.; Aquilanti, V.; De Oliveira, H.C.B.; Mundim, K.C. Uniform description of non-Arrhenius temperature dependence of reaction rates, and a heuristic criterion for quantum tunneling vs classical non-extensive distribution. *Chem. Phys. Lett.* **2013**, *590*, 201–207. [CrossRef]
60. Aquilanti, V.; Mundim, K.C.; Elango, M.; Kleijn, S.; Kasai, T. Temperature dependence of chemical and biophysical rate processes: Phenomenological approach to deviations from Arrhenius law. *Chem. Phys. Lett.* **2010**, *498*, 209–213. [CrossRef]
61. Gell-Mann, M.; Tsallis, C. *Nonextensive Entropy: Interdisciplinary Applications*; Oxford University Press: New York, NY, USA, 2004.
62. Tsallis, C. Possible generalization of Boltzman-Gibbs Statistics. *J. Stat. Phys.* **1988**, *52*, 479–487. [CrossRef]
63. Carvalho-Silva, V.H.; Aquilanti, V.; de Oliveira, H.C.B.; Mundim, K.C. Deformed transition-state theory: Deviation from Arrhenius behavior and application to bimolecular hydrogen transfer reaction rates in the tunneling regime. *J. Comput. Chem.* **2017**, *38*, 178–188. [CrossRef] [PubMed]
64. Sanches-Neto, F.O.; Coutinho, N.D.; Silva, V. A novel assessment of the role of the methyl radical and water formation channel in the CH_3OH + H reaction. *Phys. Chem. Chem. Phys.* **2017**, *19*, 24467–24477. [CrossRef] [PubMed]
65. Aquilanti, V.; Mundim, K.C.; Cavalli, S.; De Fazio, D.; Aguilar, A.; Lucas, J.M. Exact activation energies and phenomenological description of quantum tunneling for model potential energy surfaces. the F + H_2 reaction at low temperature. *Chem. Phys.* **2012**, *398*, 186–191. [CrossRef]
66. Coutinho, N.D.; Silva, V.H.C.; de Oliveira, H.C.B.; Camargo, A.J.; Mundim, K.C.; Aquilanti, V. Stereodynamical Origin of Anti-Arrhenius Kinetics: Negative Activation Energy and Roaming for a Four-Atom Reaction. *J. Phys. Chem. Lett.* **2015**, *6*, 1553–1558. [CrossRef]
67. Coutinho, N.D.; Sanches-Neto, F.O.; Carvalho-Silva, V.H.; de Oliveira, H.C.B.; Ribeiro, L.A.; Aquilanti, V. Kinetics of the OH + HCl $\leftarrow H_2O$+Cl reaction: rate determining roles of stereodynamics and roaming, and of quantum tunnelling. *J. Comput. Chem.* **2018**, *39*, 2508–2516. [CrossRef]

68. Coutinho, N.D.; Aquilanti, V.; Sanches-Neto, F.O.; Vaz, E.C.; Carvalho-Silva, V.H. First-principles molecular dynamics and computed rate constants for the series of OH-HX reactions (X= H or the halogens): non-Arrhenius kinetics, stereodynamics and quantum tunnel. In Proceedings of the 18th International Conference on Computational Science and Applications, ICCSA 2018, Melbourne, VIC, Australia, 2–5 July 2018.
69. Biró, T.; Ván, P.; Barnaföldi, G.; Ürmössy, K. Statistical Power Law due to Reservoir Fluctuations and the Universal Thermostat Independence Principle. *Entropy* **2014**, *16*, 6497–6514. [CrossRef]
70. Bíró, G.; Barnaföldi, G.; Biró, T.; Ürmössy, K.; Takács, Á.; Bíró, G.; Barnaföldi, G.G.; Biró, T.S.; Ürmössy, K.; Takács, Á. Systematic Analysis of the Non-Extensive Statistical Approach in High Energy Particle Collisions—Experiment vs. Theory. *Entropy* **2017**, *19*, 88. [CrossRef]
71. Junior, A.C.D.P.R.; Cruz, C.; Santana, W.S.; Moret, M.A. Characterization of the non-Arrhenius behavior of supercooled liquids by modeling non-additive stochastic systems. *Phys. Rev. E* **2019**, *100*, 22139.
72. Junior, A.C.D.P.R.; Cruz, C.; Santana, W.S.; Junior, E.B.A.; Moret, M.A. Non-Arrhenius behavior and fragile-to-strong transition of glass-forming liquids. *Phys. Rev. E* **2020**, *101*, 042131.
73. Bradley, R.E.; Sandifer, C.E. *Cauchy's Cours d'Analyse: An Annotated Translation*; Springer: Pasadena, CA, USA, 2009.
74. Wigner, E.P. On the Behavior of Cross Sections Near Thresholds. *Phys. Rev.* **1948**, *73*, 1002–1009. [CrossRef]
75. Takayanagi, T. *Theory of Atom Tunneling Reactions in the Gas Phase*; Springer: Berlin, Germany, 2004; pp. 15–31.
76. Sanches-Neto, F.O.; Coutinho, N.D.; Palazzetti, F.; Carvalho-Silva, V.H. Temperature dependence of rate constants for the H(D) + CH4 reaction in gas and aqueous phase: deformed Transition-State Theory study including quantum tunneling and diffusion effects. *Struct. Chem.* **2019**, *31*, 609–617. [CrossRef]
77. Abe, S.; Okamoto, Y. Nonextensive Statistical Mechanics and Its Applications. In *Lecture Notes in Physics*; Springer: Berlin, Germany, 2008.
78. Tsallis, C. *Introduction to Nonextensive Statistical Mechanics: Approaching a Complex World*; Springer: New York, NY, USA, 2009.
79. Tolman, R.C. Statistical Merchanics Applied To Chemical Kinetics. *J. Am. Chem. Soc.* **1920**, *42*, 2506–2528. [CrossRef]
80. Vogel, H. Das temperature-abhangigketsgesetz der viskositat von flussigkeiten. *Phys. Z* **1921**, *22*, 645–646.
81. Fulcher, G.S. Analysis of Recent Measurements of the Viscosity of Glasses. *J. Am. Ceram. Soc.* **1925**, *8*, 339–355. [CrossRef]
82. Tammann, G.; Hesse, W. Die Abhängigkeit der Viscosität von der Temperatur bie unterkühlten Flüssigkeiten. *Zeitschrift für Anorg. und Allg. Chemie* **1926**, *156*, 245–257. [CrossRef]
83. Laidler, K.J. A Glossary of Terms Used in Chemical Kinetics, Including Reaction Dynamics. *Pure Appl. Chem.* **1996**, *68*, 149–192. [CrossRef]
84. Planck, M.; Press, P. On the Theory of the Energy Distribution Law of the Normal Spectrum. *Verh. Deutsch. Phys. Ges.* **1900**, *237*, 1–8.
85. Slater, N.B. *Theory of Unimolecular Reactions*; Cornell University Press: Ithaca, NY, USA, 1959.
86. Marcelin, R. Contribution à l'étude de la cinétique physico-chimique. *Ann. Phys. (Paris).* **1915**, *9*, 120–231. [CrossRef]
87. Callan, C.G. Broken scale invariance in scalar field theory. *Phys. Rev. D* **1970**, *2*, 1541. [CrossRef]
88. Symanzik, K. Small-distance-behaviour analysis and Wilson expansions. *Commun. Math. Phys.* **1971**, *23*, 49–86. [CrossRef]
89. Wilson, K.G. Renormalization group and critical phenomena. I. Renormalization group and the Kadanoff scaling picture. *Phys. Rev. B* **1971**, *4*, 3174. [CrossRef]
90. Weinberg, S. *The Quantum Theory of Fields*; Cambridge University Press: Cambridge, UK, 1995; Volume 2.
91. Gupta, S. Effective Field Theories. In Lecture 2: Wilsonian Renormalization. In Proceedings of the Mini School 2016, IACS Kolkata, India, 29 February–4 March 2016.
92. Honig, J.M.; Spalek, J. *A Primer to the Theory of Critical Phenomena*; Elsevier Science: Amsterdam, The Netherlands, 2018.
93. Mallamace, F.; Branca, C.; Corsaro, C.; Leone, N.; Spooren, J.; Chen, S.H.; Stanley, H.E. Transport properties of glass-forming liquids suggest that dynamic crossover temperature is as important as the glass transition temperature. *Proc. Natl. Acad. Sci. USA* **2010**, *107*, 22457–22462. [CrossRef] [PubMed]

94. Martinez-Garcia, J.C.; Martinez-Garcia, J.; Rzoska, S.J.; Hulliger, J. The new insight into dynamic crossover in glass forming liquids from the apparent enthalpy analysis. *J. Chem. Phys.* **2012**, *137*, 064501. [CrossRef] [PubMed]
95. Souletie, J. The glass transition: dynamic and static scaling approach. *J. Phys.* **1990**, *51*, 883–898. [CrossRef]
96. Mauro, J.C.; Yue, Y.; Ellison, A.J.; Gupta, P.K.; Allan, D.C. Viscosity of glass-forming liquids. *Proc. Natl. Acad. Sci. USA* **2009**, *106*, 19780–19784. [CrossRef]
97. Souletie, J.; Tholence, J.L. Critical slowing down in spin glasses and other glasses: Fulcher versus power law. *Phys. Rev. B* **1985**, *32*, 516. [CrossRef]
98. Stickel, F.; Fischer, E.W.; Richert, R. Dynamics of glass-forming liquids. II. Detailed comparison of dielectric relaxation, de-conductivity, and viscosity data. *J. Chem. Phys.* **1996**, *104*, 2043. [CrossRef]
99. Drozd-Rzoska, A. Universal behavior of the apparent fragility in ultraslow glass forming systems. *Sci. Rep.* **2019**, *9*, 6816. [CrossRef]
100. Coutinho, N.D.; Silva, V.H.C.; Mundim, K.C.; de Oliveira, H.C.B. Description of the effect of temperature on food systems using the deformed Arrhenius rate law: deviations from linearity in logarithmic plots vs. inverse temperature. *Rend. Fis. Acc. Lincei* **2015**, *26*, 141–149. [CrossRef]
101. Cavalli, S.; Aquilanti, V.; Mundim, K.C.; De Fazio, D. Theoretical reaction kinetics astride the transition between moderate and deep tunneling regimes: The F + HD case. *J. Phys. Chem. A* **2014**, *118*, 6632–6641. [CrossRef]
102. Crozier, W.J.; Stier, T.B. Critical Thermal Increments for Rhythmic Respiratory Movements of Insects. *J. Gen. Physiol.* **1925**, *7*, 429–447. [CrossRef] [PubMed]
103. Cossins, A.R.; Bowler, K. *Temperature Biology of Animals*; Chapman and Hall: London, UK, 1987.
104. Giordano, D.; Dingwell, D.B. Non-Arrhenian multicomponent melt viscosity: A model. *Earth Planet. Sci. Lett.* **2003**, *208*, 337–349. [CrossRef]
105. Giordano, D.; Mangiacapra, A.; Potuzak, M.; Russell, J.K.; Romano, C.; Dingwell, D.B.; Di Muro, A. An expanded non-Arrhenian model for silicate melt viscosity: A treatment for metaluminous, peraluminous and peralkaline liquids. *Chem. Geol.* **2006**, *229*, 42–56. [CrossRef]
106. Suleimanov, Y. V.; Collepardo-Guevara, R.; Manolopoulos, D.E. Bimolecular reaction rates from ring polymer molecular dynamics: Application to $H + CH_4 \rightarrow H_2 + CH_3$. *J. Chem. Phys.* **2011**, *134*, 044131. [CrossRef]
107. Zhang, Y.; Stecher, T.; Cvitaš, M.T.; Althorpe, S.C. Which Is Better at Predicting Quantum-Tunneling Rates: Quantum Transition-State Theory or Free-Energy Instanton Theory? *J. Phys. Chem. Lett.* **2014**, *5*, 3976–3980. [CrossRef]
108. Richardson, J.O. *Ring-Polymer Approaches to Instanton Theory*; University of Cambridge: London, UK, 2012.
109. Janssen, L.M.C. Mode-coupling theory of the glass transition: A primer. *Front. Phys.* **2018**, *6*, 97. [CrossRef]
110. Müller, I.; Ruggeri, T. *Rational Extended Thermodynamics*; Springer: New York, NY, USA, 1998.
111. Williams, M.L.; Landel, R.F.; Ferry, J.D. Temperature Dependence of Relaxation Mechanisms in Amorphous Polymers and Other Glass-Forming Liquids. *Phys. Rev.* **1955**, *98*, 1549. [CrossRef]
112. Bělehrádek, J. A unified theory of cellular rate processes based upon an analysis of temperature action. *Protoplasma* **1957**, *48*, 53–71. [CrossRef]
113. Bässler, H. Viscous flow in supercooled liquids analyzed in terms of transport theory for random media with energetic disorder. *Phys. Rev. Lett.* **1987**, *58*, 767–770. [CrossRef]
114. Nakamura, K.; Takayanagi, T.; Sato, S. A modified arrhenius equation. *Chem. Phys. Lett.* **1989**, *160*, 295–298. [CrossRef]
115. Gentili, P.L. The fuzziness of the molecular world and its perspectives. *Molecules* **2018**, *23*, 2074. [CrossRef] [PubMed]
116. Zadeh, L.A. Toward human level machine intelligence - Is it achievable? the need for a paradigm shift. *IEEE Comput. Intell. Mag.* **2008**, *3*, 11–22. [CrossRef]
117. Duffin, R.J.; Zener, C. Geometric programming and the Darwin-Fowler method in statistical mechanics. *J. Phys. Chem.* **1970**, *74*, 2419–2423. [CrossRef]
118. Agrawal, A.; Diamond, S.; Boyd, S. Disciplined geometric programming. *Optim. Lett.* **2019**, *13*, 961–976. [CrossRef]
119. Kamel, A.M.; Dorrah, H.T.; Farid, S.; Eid, S.; Mohammed, A. Fuzzy Geometric Programming Optimization using New Arithmetic Fuzzy Logic-based Representation. In Proceedings of the 14thInternational Middle East PowerSystems Conference (MEPCON'10) , Cairo, Egypt, 19–21 December 2010.

120. Mundim, K.C.; Tsallis, C. Geometry optimization and conformational analysis through generalized simulated annealing. *Int. J. Quantum Chem.* **1996**, *58*, 373–381. [CrossRef]
121. Ilva, R. A Maxwellian Path to the q -Nonextensive Velocity Distribution Function. *Phys. Lett. A* **2008**, *9601*, 1–15.
122. Silva, R.; Alcaniz, J.S. Negative heat capacity and non-extensive kinetic theory. *Phys. Lett. Sect. A Gen. At. Solid State Phys.* **2003**, *313*, 393–396. [CrossRef]
123. Laidler, K. Chemical Kinetics. *Harper Row* **1987**, *1*, 233–270.

molecules

MDPI

Article

"Transitivity": A Code for Computing Kinetic and Related Parameters in Chemical Transformations and Transport Phenomena

Hugo G. Machado [1,2,*], Flávio O. Sanches-Neto [1,2,*], Nayara D. Coutinho [3,*], Kleber C. Mundim [2], Federico Palazzetti [3,*] and Valter H. Carvalho-Silva [1,2,*]

1 Grupo de Química Teórica e Estrutural de Anápolis, Centro de Pesquisa e Pós-Graduação, Universidade Estadual de Goiás, 75132-400 Anápolis, GO, Brazil

2 Instituto de Química, Universidade de Brasília, Caixa Postal 4478, 70904-970 Brasília, Brazil; kcmundim@unb.br

3 Dipartimento di Chimica, Biologia e Biotecnologie, Università di Perugia, 06123 Perugia, Italy

* Correspondence: hugogontijomachado@gmail.com (H.G.M.); flavio_olimpio@outlook.com (F.O.S.-N.); nayaradcoutinho@gmail.com (N.D.C.); fede_75it@yahoo.it (F.P.); fatioleg@gmail.com (V.H.C.-S.); Tel.: +55-62-3328-1156 (V.H.C.-S)

Academic Editor: Pier Luigi Gentili

Received: 16 August 2019; Accepted: 8 September 2019; Published: 25 September 2019

Abstract: The Transitivity function, defined in terms of the reciprocal of the apparent activation energy, measures the propensity for a reaction to proceed and can provide a tool for implementing phenomenological kinetic models. Applications to systems which deviate from the Arrhenius law at low temperature encouraged the development of a user-friendly graphical interface for estimating the kinetic and thermodynamic parameters of physical and chemical processes. Here, we document the Transitivity code, written in Python, a free open-source code compatible with Windows, Linux and macOS platforms. Procedures are made available to evaluate the phenomenology of the temperature dependence of rate constants for processes from the Arrhenius and Transitivity plots. Reaction rate constants can be calculated by the traditional Transition-State Theory using a set of one-dimensional tunneling corrections (Bell (1935), Bell (1958), Skodje and Truhlar and, in particular, the deformed (*d*-TST) approach). To account for the solvent effect on reaction rate constant, implementation is given of the Kramers and of Collins–Kimball formulations. An input file generator is provided to run various molecular dynamics approaches in CPMD code. Examples are worked out and made available for testing. The novelty of this code is its general scope and particular exploit of *d*-formulations to cope with non-Arrhenius behavior at low temperatures, a topic which is the focus of recent intense investigations. We expect that this code serves as a quick and practical tool for data documentation from electronic structure calculations: It presents a very intuitive graphical interface which we believe to provide an excellent working tool for researchers and as courseware to teach statistical thermodynamics, thermochemistry, kinetics, and related areas.

Keywords: *d*-TST; activation energy; Transitivity plot; solution kinetic

1. Introduction

Recent applications of chemical kinetics to a variety of complex systems involves the accurate dealing of properties to be described by techniques, which treat a series of processes beyond elementary chemical quantum dynamics or even approximate classical and semiclassical approaches. We analyze in a companion paper [1] the state of our approaches to these problems from a general viewpoint. Here, we deal with explicit computational calculations that allow moving directly in a simple way to global applications.

Recourse needs to be made at one stage or another to statistical treatments [1–3] among which there is modern insurgence with respect to more traditional ones—exemplary is, in particular, this topical collection essentially dedicated to "Fuzzy Logic" [4,5].

Information on the kinetic and related parameters in chemical transformations and transport phenomena and their role in complex mechanisms is needed: Particularly the temperature dependence of rate processes, $k(T)$ and often in the low temperature range, where deviations from linearity of Arrhenius plots are revealed. The phenomenology of curvatures in Arrhenius plot span all of chemistry: From the long list that is continuously updated, we refer here to some selected cases, such as combustion chemistry [6], condensed-phase [7], atmospheric and astrochemical reactions [8,9], processes involved in the preservation and aging of food and drugs [10,11], as well as in basic geochemical [12] and biochemical environments [13,14]. The current status of the phenomenology is classified emphasizing case studies, specifically (i) super-Arrhenius kinetics, convex curvature in the Arrhenius plot, where transport phenomena brakes the processes as temperature decreases; (ii) sub-Arrhenius kinetics, concave curvature in the Arrhenius plot, where quantum mechanical tunneling propitiates low temperature reactivity; (iii) anti-Arrhenius kinetics, negative apparent activation energy, where processes are limited by stereodynamic requirements.

The curvature in the Arrhenius plot promotes a temperature dependence on the apparent activation energy E_a, the definition of which has been recommended by the International Union for Pure and Applied Chemistry (IUPAC) [13,15] as follows,

$$E_a = -k_B \frac{\mathrm{d} \ln k(T)}{\mathrm{d}\left(\frac{1}{T}\right)} = -\frac{\mathrm{d} \ln k(\beta)}{\mathrm{d}\,\beta}. \tag{1}$$

In Equation (1), k_B is the Boltzmann constant, T is the absolute temperature and $\beta = 1/k_BT$. The so-called Tolman's Theorem [16] proposed an interpretation for E_a as the difference between the average over the energy of all reacting systems and energy of all (reactive and nonreactive) systems—propitiating a connection between canonical quantities with microcanonical features of the potential energy surface. Independently, the same spirit lead to transition-state formulations [15–18]. Recently, it was found expedient to introduce the reciprocal of the apparent activation energy in order to define the Transitivity function [2,3], Equation (2),

$$\gamma(\beta) \equiv \frac{1}{E_a(\beta)} \tag{2}$$

to construct an appropriate scaling plane, $\gamma(\beta)$ vs. β, where the regular curvatures on the Arrhenius plane are approximately linearized: The Transitivity plot. The function $\gamma(\beta)$ can be interpreted as a measure of the propensity for the reaction to proceed and permit uniformly to account for experimental and theoretical rate processes, such as quantum tunneling, transport properties, and diffusion in the neighborhoods of phase transitions.

For cases where experimental and theoretical rate processes are difficult to interpret—by molecular complexity, extreme conditions or impossibility of exact solution to the Schrodinger's equation—phenomenological and semiclassical theoretical approaches are of increasing utility. Over the years, the paramount useful phenomenological tool for studying the kinetics of physicochemical processes has been the Arrhenius law. It fails when the temperature range of the rate process becomes large: Empirical laws have been proposed, involving empirical parameters often lacking of physical interpretation, e.g., Kooij [19], power law [20], Vogel–Fulcher–Tamman (VFT) [21–23], Nakamura–Takayanagi–Sato (NTS) [24], deformed (*d*) Aquilanti–Mundim (AM) [25] and Aquilanti–Sanchez–Coutinho–Carvalho (ASCC) [26]. In Reference [27], it was presented how the transitivity concept can sort and interpret all these empirical laws, permitting a microscopic interpretation of the phenomenological parameters.

For understanding and predicting a wide variety of kinetic processes with large molecular complexity and presenting deviation from Arrhenius law, Transition-State Theory (TST) remains an excellent formulation. The TST triggered the development of a variety of improved approaches: Important variants include the variational TST [28], the Marcus theory of electron transfer [29], and quantum [30] and path integral [31,32] versions. However, the variants of TST require additional information of the potential energy surface rendering formidable the task of estimating kinetic parameters as the size of molecular systems increases: Examples are the degradation kinetics of organic pollutants in aquatic, soil, and atmospheric environments [33–38].

The late twentieth century saw the advent, beyond the Transition State formulation, of molecular dynamics simulations, which is nowadays a powerful theoretical tool in understanding mechanisms of physical and chemical processes. Recently, there is a vast activity evaluating whether advances in molecular dynamics simulations can provide quantitatively rate constants [39–41]. However, the extraction of quantitative information on rate constants from molecular dynamics simulations is an important issue but a very difficult one to tackle [42–44].

Software tools to calculate and interpret rate constants are enormously useful in material, biochemical, and geochemical research to permit the exploitation of intense progress in computational hardware. Widely diverse codes have been developed to calculate rate constants in a gas-, liquid-, and solid-phase using Transition-State Theory and its variants: Polyrate [45], TheRate [46], MultiWell [47], TAMkin [48], Mesmer [49], RMG [50], APUAMA [51], KiSThelP [52], FRIGUS [53] and Eyringpy [54] codes are excellent options available to work out the kinetics of chemical reactions. Herein, we describe a new code with a user-friendly graphical interface able to perform various procedures for computing kinetic and related parameters in chemical transformations and transport phenomena: The code is referred to as "Transitivity" (see Figure 1), a concept related to the function defined in Equation (2); written in Python, it is a free open-source code compatible with Windows, Linux, and MacOS platforms. It offers the possibility of estimating phenomenological parameters from Arrhenius and Transitivity plots using a stochastic optimization method, Generalized Simulated Annealing (GSA) [55], with several options: Arrhenius, Aquilanti–Mundim (AM) [25], Vogel–Fulcher–Tammann (VFT) [22], Nakamura–Takayanagi–Sato (NTS) [24,56] and Aquilanti–Sanchez–Coutinho–Carvalho (ASCC) [26] formulas. This code also allows the estimation of unimolecular and bimolecular reaction rate constants with traditional TST using Bell35 [57], Bell58 [58], Skodje–Truhlar (ST) [59] and deformed (*d*-TST) corrections [60]. Solvent effects can be accounted for by the Collins–Kimball [61] and Kramers [62] models. Finally, input files for different first-principles molecular dynamics—Born–Oppenheimer Molecular Dynamics (BOMD), Car–Parrinello Molecular Dynamics (CPMD), Metadynamics, Path-Integral Molecular Dynamics (PIMD), and Trajectory Surface Hopping (TSH)—can be generated to run by the CPMD code [63].

We propose this code, as we expect it serves as a quick and practical tool for documentation data from electronic structure calculations. Additionally, it presents a very intuitive graphical interface which we believe to provide a useful working tool for the general public and researchers and also as courseware to teach statistical thermodynamics, thermochemistry, kinetics, and related areas. The body of this article is structured in two parts, one that deals with a brief theoretical and operational description of the code, and a second one that shows a variety of examples. A final section is devoted to additional and concluding remarks and an Appendix A collects used symbols and their meaning.

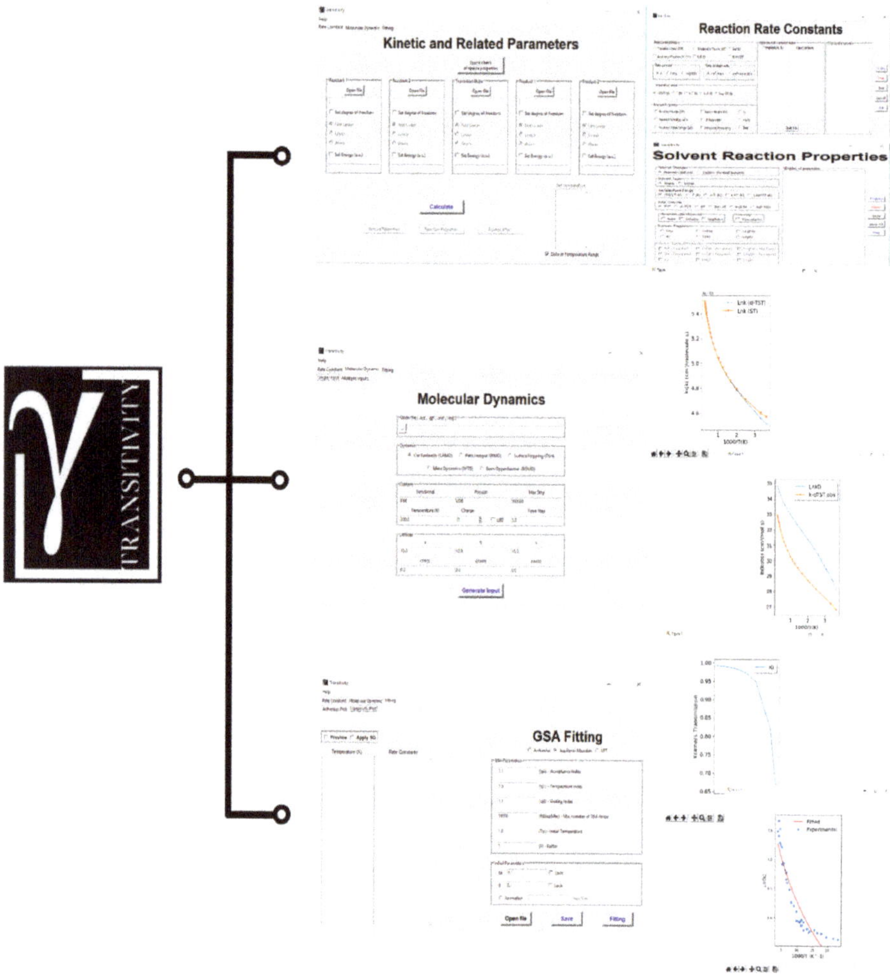

Figure 1. Logo and main windows of the Transitivity code.

2. Theoretical Background

2.1. Phenomenology of Temperature Dependence of the Reaction Rate Constant

The theoretical apparatus to connect the Transitivity function, $\gamma(T)$, and phenomenological reaction rate constant formulas and vice-versa is built and extensively discussed in Reference [27] and references therein. Classical and recent phenomenological reaction rate constant formulas to account for sub-, super-, and anti-Arrhenius behavior are applied in Section 4.1 by use of: (i) Deformed (*d*) Aquilanti–Mundim [25], (ii) VFT (Vogel–Fulcher–Tammann) [22], (iii) NTS (Nakamura–Takayanagi–Sato) [24,56], and (iv) ASCC (Aquilanti–Sanchez–Coutinho–Carvalho) [26]. More details of the formulas are presented in Table 1.

2.2. Calculation of Reaction Rate Constant

The Transition-State Theory, TST, is the most popular tool used to study the kinetics of chemical reactions with a well-defined activated complex—as customary, a double dagger (‡) denotes the properties pertaining to the transition-state complex. For a general bimolecular reaction, such as Reactants $\rightarrow TS^{\ddagger} \rightarrow$ Products, it is necessary to compute the Q_1, Q_2 and $Q^{\ddagger}$ partition functions of reactants R_1 and R_2 and of the transition state, respectively. At absolute temperature T, the rate constant is given by:

$$k^{TST}(T) = \frac{k_B T}{h} \frac{Q^{\ddagger}}{Q_1 Q_2} \exp\left(-\frac{\varepsilon^{\ddagger}}{k_B T}\right), \tag{3}$$

where h is Planck's constant; and $\varepsilon^{\ddagger}$ is the effective height of the energy barrier, eventually with the addition of the zero-point energy correction in Equation (3). The tunneling correction introduced in TST can be calculated by Skodje–Truhlar [59], Bell35 [57], and Bell58 [58] corrections and by the deformed (d) tunneling formulation presented in the next section [60].

2.2.1. Deformed Transition-State Theory (d-TST)

The deformed Transition-State Theory (d-TST) [60] formulation:

$$k_d(T) = \frac{k_B T}{h} \frac{Q^{\ddagger}}{Q_1 Q_2} \left(1 - d\frac{\varepsilon^{\ddagger}}{k_B T}\right)^{1/d}, \quad d = -\frac{1}{3}\left(\frac{h\nu^{\ddagger}}{2\varepsilon^{\ddagger}}\right)^2, \tag{4}$$

where $\nu^{\ddagger}$ is frequency for crossing the barrier, uniformly covers the range from classical to moderate tunneling regimes but needs an amendment for deep tunneling in exothermic reactions, a relatively rare case. The proposed variant of the transition-state theory was obtained from the transitivity concept and deformed (d) Aquilanti–Mundim law permit comparison with experiments and tests against alternative formulations. The nomenclature in Equation (4) is the same as used in Equation (3). A popular formulation in the literature is to use the Wigner tunneling correction: However, in our previous study [60,64], we have shown that d-TST is a more satisfactory approximation also in view of uniform behavior across the height of the energy barrier.

2.2.2. Bell35 and Bell58

To cover the moderate-to-deep tunneling transition in exothermic reactions, we applied both Bell35 correction [57,65], Equation (5):

$$\kappa_{Bell35} = \frac{\left[\frac{1}{\hbar\nu^{\ddagger}} - \frac{1}{k_B T}\exp\left(\frac{\varepsilon^{\ddagger}}{k_B T} - \frac{\varepsilon^{\ddagger}}{\hbar\nu^{\ddagger}}\right)\right]}{\frac{1}{\hbar\nu^{\ddagger}} - \frac{1}{k_B T}}, \tag{5}$$

and Bell58 [58,65] correction truncated at the second term ($2T$)

$$\kappa_{Bell58-2T} = \frac{\left(\frac{h\nu^{\ddagger}}{2k_B T}\right)}{\sin\left(\frac{h\nu^{\ddagger}}{2k_B T}\right)} - \frac{\exp\left(\frac{\varepsilon^{\ddagger}}{k_B T} - \frac{\varepsilon^{\ddagger}}{\hbar\nu^{\ddagger}}\right)}{\left(\frac{k_B T}{\hbar\nu^{\ddagger}} - 1\right)}, \tag{6}$$

that, although non-uniform across the transition between negligible and moderate tunneling regimes, were found to behave smoothly enough to adequately perform practically astride the whole range.

Truncating Equation (6) to the first term, we recover the usual formula used in the literature to describe the tunneling under deep regime Equation (7):

$$\kappa_{Bell58} = \frac{\left(\frac{h\nu^{\ddagger}}{2k_B T}\right)}{\sin\left(\frac{h\nu^{\ddagger}}{2k_B T}\right)} \tag{7}$$

However, this formulation presents divergence at the crossover temperature $T_c = h\nu^{\ddagger}/\pi k_B$. The tunneling regimes can be delimited within four temperatures ranges—negligible ($T > 2T_c$), small ($2T_c > T > T_c$), moderate ($T_c > T > T_c/2$), and deep ($T > T_c/2$) [60,66,67].

2.2.3. Skodje and Truhlar, ST

To avoid spurious divergence at T_c in the Bell58 formulation, in 1981, Skodje and Truhlar [59] gave a generalization extending the parabolic barrier treatment. In their approximation for tunneling correction Equations (8a) and (8b)

$$\kappa_{ST} = \frac{\left(\frac{h\nu^{\ddagger}}{2k_BT}\right)}{\sin\left(\frac{h\nu^{\ddagger}}{2k_BT}\right)} - \frac{\exp\left[\left(\frac{1}{k_BT} - \frac{1}{\hbar\nu^{\ddagger}}\right)\left(\varepsilon^{\ddagger} - \Delta H\right)\right]}{\left(\frac{k_BT}{\hbar\nu^{\ddagger}} - 1\right)}, \quad \beta \leq \mathbb{C}, \tag{8a}$$

and

$$\kappa_{ST} = \frac{1}{\left(\frac{k_BT}{\hbar\nu^{\ddagger}} - 1\right)}\left\{\exp\left[\left(\frac{1}{k_BT} - \frac{1}{\hbar\nu^{\ddagger}}\right)\left(\varepsilon^{\ddagger} - \Delta H\right)\right] - 1\right\}, \quad \mathbb{C} \leq \beta, \tag{8b}$$

where $\mathbb{C} = 1/k_BT_c = \pi/h\nu^{\ddagger}$ and ΔH is the enthalpy of reaction.

2.3. *Solvent Effect on Reaction Rate Constant*

2.3.1. Collins–Kimball Formulation

Treatment of chemical reactions in liquid-phase requires accounting for the solvent effect, considering the ability of the reagents to diffuse and to lead them to effective reactive collisions. According to the Onsager solvent reaction field model [68–70], the solvent creates a solvation layer, as a cage around the molecular entities that participate in the reaction. The reactive process between molecular entities A and B is represented by the sequential equation,

$$A_{solv} + B_{solv} \underset{\overleftarrow{k}_D}{\overset{\overrightarrow{k}_D}{\rightleftharpoons}} \{AB\}_{solv} \xrightarrow{k} Product. \tag{9}$$

The *"solv"* symbol indicates that the molecular entity is surrounded by a roughly spherical solvation layer, the cage, having a specific radius. $\overrightarrow{k}_D$ is the diffusion kinetic rate constant which the reagent A travels in the solvent to find B; vice versa, $\overleftarrow{k}_D$ is the kinetic rate constant for reverse diffusion; k denotes the reaction rate constant due to effective reactive collisions: It can be estimated from Equations (3) or (4). Assuming the Steady-State Approximation [70] for Equation (9), the Smoluchowski expression for the diffusion kinetic rate constant [71], takes the form Equation (10):

$$\overrightarrow{k}_D = 4\pi r_{AB} D_{AB}, \tag{10}$$

where r_{AB} is a reaction radial distance and D_{AB} is the sum of the diffusion constants for each reagent in the solvent. The generalization of Collins and Kimball [61,72] for the irreversible bimolecular diffusion-controlled reactions at infinite reaction rate, $\overrightarrow{k}_D \cong \overleftarrow{k}_D$ [73], yields an overall reaction rate constant, k_{Obs} Equation (11)

$$\frac{1}{k_{Obs}} = \frac{1}{k_{TST}} + \frac{1}{\overrightarrow{k}_D}. \tag{11}$$

In the code, the diffusion is accounted for by the Stokes–Einstein formulation and the temperature dependence of viscosity η of the solvent is estimated through the Aquilanti–Mundim formula (see details in Section 3).

2.3.2. Kramers' Formulation

To account for dynamical effects of the solvent in a reactive process and to generalize to unimolecular and pseudo-unimolecular processes, Kramers' model considers a stochastic motion of the system, where the solvent effect is added considering Brownian movements along the reaction path [70].

Assuming that the friction constant, μ (see below), is independent of time, the overall reaction rate constant k_{Obs} can be calculated as Equation (12)

$$k_{Obs} = \kappa_{Kr} k_{TST}, \tag{12}$$

where κ_{Kr} is the transmission factor obtained by Kramers [62] as:

$$\kappa_{Kr} = \frac{1}{\omega^{\ddagger}}\left(\sqrt{\frac{\mu^2}{4} + \omega^{\ddagger 2}} - \frac{\mu}{2}\right). \tag{13}$$

For the transition-state theory rate constant k_{TST} and variants see Section 2.2. In Equation (13), $\omega^{\ddagger}$ is the imaginary frequency of the transition state and the friction constant is given by $\mu = (6\pi r_{AB}/M)\eta$, where r_{AB} and M are the radius of the cage and the molecular mass of the transition-state, respectively. Again, the viscosity is calculated in the code through the Aquilanti–Mundim formula.

3. Handling the Transitivity Code

Several aspects regarding nomenclature and the theoretical background used in this article are presented in already cited references and in Reference [27]. Furthermore, the code with the manual, examples (see next section), and installation video can be freely downloaded in the www.vhcsgroup.com/transitivity web page.

In the main window of the code, the user can choose between three options: (i) Fitting reaction rate constant data as a function of temperature; (ii) predicting the reaction rate constant in the gas- and liquid-phase; and (iii) creating the input for the calculation of first-principles molecular dynamics by the CPMD code. If the option chosen is "Kinetic and Related Parameters", the program needs the electronic structure output files provided by the Gaussian program for the structures of the reagents, of the transition state and of the products. If the system under study requires the calculation of the electronic energy using a specifically high level method, the energy values must be included separately in the box. The code provides the reaction rate constants for both unimolecular and bimolecular reactions and the users should indicate if the molecular entity corresponds to an atom or whether the molecule is linear or not. In a new window, the user can choose between TST or *d*-TST [60] and Bell35 [57], Bell58 [58], Skodje–Truhlar (ST) [59] tunneling corrections. In addition, the program provides a visualization of the Arrhenius plot with the possibility of including any experimental and/or theoretical data available for validation. In the same window, reaction properties (such as internal energy, enthalpy, Gibbs-free energy, barrier height, *d* parameter, imaginary frequency ($\nu^{\ddagger}$), crossover temperature (T_c), and the parameter of the Skodje–Truhlar model) are exhibited.

If the VOLUME keyword is used in the calculation input file of G09, the option including solvent effects is available through the Kramers and Collins–Kimball formulations. It is necessary to make the choice of solvent parameters for the estimation of the viscosity or use the default (water). The parameters for the viscosity estimation ($\eta/Poise$) of a solvent other than water should be those fitted by the Aquilanti–Mundim formula [25] with η_o in Poise and ε in J·mol^{-1} and inserted in the "Solvent Type" option. The temperature dependence of viscosity of water using the experimental data [74] is expressed as $\eta(T) = 2.7024.10^{-4} Poise(1 - 213.0543/T)^{-2.75634}$. When the Kramers formulation is chosen, the friction coefficient of the solvent (μ/s^{-1}), the Kramers transmission coefficient, and the overall reaction rate constant (k_{Obs}/cm^3·mol^{-1}·s^{-1} or s^{-1}) are provided. If the Collins–Kimball formulation is selected, the overall reaction rate constant (k_{Obs}/cm^3·mol^{-1}.s^{-1}), separate diffusion coefficients (cm^2·s^{-1}) for the reactants and the Smoluchowski diffusion rate constant ($\vec{k}_D$/cm^3·mol^{-1}·s^{-1}) are provided.

With the "GSA Fitting" option, the values of the rate constants and temperatures will be needed. In the software, the fitting in the Arrhenius plot is implemented for the rate constant data using the Arrhenius, Aquilanti–Mundim, VFT, NTS, and ASCC formulas by the stochastic optimization algorithm GSA. It is possible to insert the guess parameters d, $\varepsilon^{\ddagger}$, E_a, E_0, E_ν, B and T_0 (see formulas in Table 1). Furthermore, information of the fit are available in "Fit [FormulaName].dat" output file. Internal parameters of GSA can be also controlled [55]. The fitting in the Transitivity plot is implemented of the reaction rate constant data using only Arrhenius, Aquilanti–Mundim, and VFT. The Transitivity plot is calculated using numerical differentiation with the option "Preview" and if necessary the smoothing of the data can be applied with the option "Apply SG" enabling the Savitzky–Golay filter [75]. In addition, the program provides a visualization of the Arrhenius and Transitivity plots with the fitting model chosen step-by-step.

The present code also offers the possibility of creating input to first-principles molecular dynamics simulation by the CPMD computational code. After selecting the input file of the system to be studied, which contains the molecular geometry of the system in a specific extension (*.xyz, *.gjf, *.out and *.log), the user should choose the molecular dynamics method. In addition, the user has the option to choose the Density Functional Theory (DFT) functional, pseudopotential, the temperature, the charge, the simulation time, and the integration time step. The size of the simulation box can be changed in the "Lattices" section. Additionally, the Transitivity code generates another output file with a *.gjf extension, where the user can check if the geometry of the system to be simulated is correct.

4. Examples

4.1. Fitting Mode—Arrhenius and Transitivity Plots

To illustrate the use of most of functionalities and to validate the accuracy of the running of the Transitivity code, a fitting of the reaction rate constants in the Arrhenius plot is performed as a function of temperature for four different systems in different regimes: sub-Arrhenius, corresponding to both deep and moderate tunneling, super-Arrhenius, and anti-Arrhenius behavior.

The first example concerns the keto–enol tautomerization of 2-(2′-hydroxy-4′-methylphenyl) benzoxazole (MeBO), a well-known case of a deep-tunneling regime [7]: For this case we performed a fitting using Arrhenius, Aquilanti–Mundim, Nakamura–Takayanagi–Sato and ASCC formulas. In order to present the results for the sub-Arrhenius behavior under the moderate-tunneling regime, we evaluated the $OH + H_2 \rightarrow H + H_2O$ reaction [76] using the Arrhenius and Aquilanti–Mundim formulas. The third example is centered on the investigations that have revealed the super-Arrhenius behavior for the rates of the processes promoted by enzymatic catalysis [77–79]. Here, we regard the reaction rate constant of the hydride transfer between the substrate and NAD^+, catalyzed by F147L, which exhibits a strong convex curvature in the temperature range of 5 to 65 °C [78]: A fitting is performed with the Arrhenius, Aquilanti–Mundim and VTF formulas. The final case is aimed at showing a case of anti-Arrhenius behavior, which is characterized by the decrease of the reaction rate constant with the increase in the temperature. The $OH + HBr \rightarrow Br + H_2O$ reaction is prototypical in studies of an example, both from a theoretical and an experimental point of view: It exhibits negative temperature dependence of the reaction rate constant [80–82]. Fitting is performed with the Arrhenius and the Aquilanti–Mundim formulas.

The fitting parameters and statistical analysis of the quality of the fits for all the systems are given in Table 1. The statistical measure χ^2 is used in tests to compare the quality of the formula with the reference. It is noteworthy that in a fitting process, in order to obtain parameters with a physical meaning, it is necessary to avoid compensation effects [83] exploiting any prior knowledge of the system, since the formulas with more than two parameters can lead to multiple solutions.

Figure 2 shows the experimental data and fitted formulas for each system in the Arrhenius plot. Both NTS and ASCC formulas were satisfactory for fitting sub-Arrhenius behavior under deep-tunneling regime, while Aquilanti–Mundim is satisfactory just within a specific range of temperature. Under moderate-tunneling regime, the Aquilanti–Mundim formula was an excellent option to describe experimental data for $OH + H_2$ reaction. The VTF and Aquilanti–Mundim formulas seem adequate to fit super-Arrhenius behavior. The Aquilanti–Mundim parameters that were obtained for the OH + HBr reaction indicate that they provide an excellent option for the anti-Arrhenius behavior. As expected, the Arrhenius formula is clearly inadequate to account for deviations at low temperature for all the reactions presented.

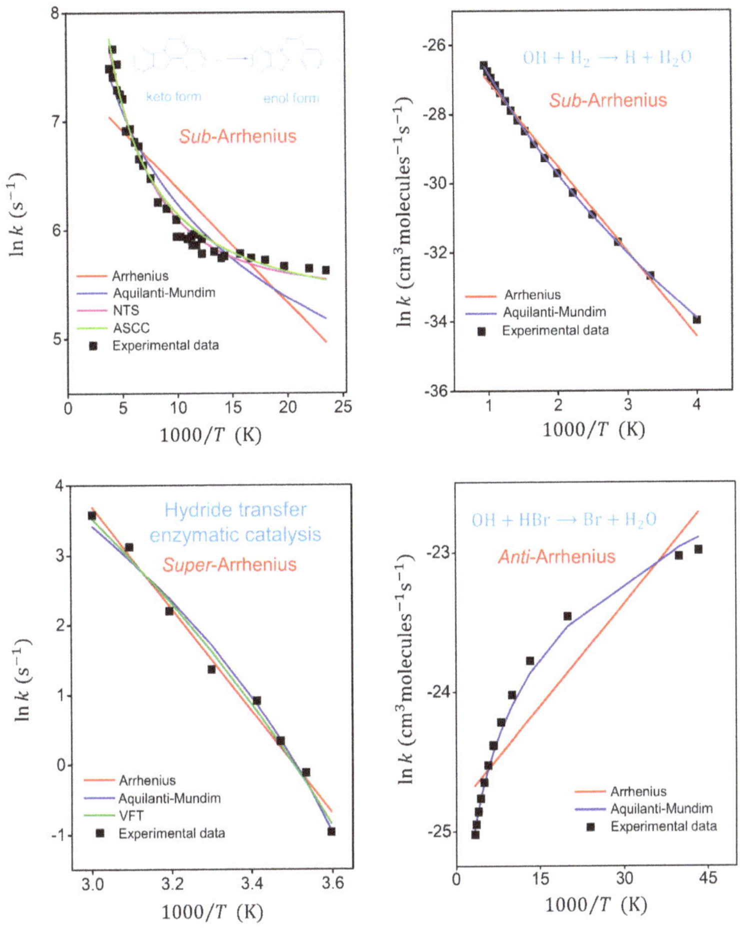

Figure 2. Arrhenius plots comparing the experimental reaction rate constant and fitted formulas for keto–enol tautomerization reaction (sub-Arrhenius behavior under deep tunneling), OH + $H_2 \longrightarrow H_2O$ + H reaction (sub-Arrhenius behavior under moderate tunneling), hydride transfer with enzymatic catalysis (super-Arrhenius behavior) and OH + HBr $\longrightarrow H_2O$ + Br reaction (anti-Arrhenius behavior). NTS and ASCC formulas were of use for sub-Arrhenius behavior under deep-tunneling regime. The Aquilanti–Mundim formula was of use for sub-Arrhenius cases under moderate-tunneling regime, for super-Arrhenius and for anti-Arrhenius behaviors. VFT also was of use for super-Arrhenius situations. The references of experimental data can be found in Table 1.

Table 1. Fitted parameters for the Arrhenius, AM, ASCC, NTS and VFT formulas, using the Transitivity code for keto–enol tautomerization [7], OH + H_2 [76], enzymatic catalysis [78] and OH + Br [84] reactions. Energy (E_a, $\varepsilon^\ddagger$, E_v and E_0) is in cal/mol and temperature (T_0 and B) in K. Pre-factor units can be identified in the references.

Formula	Fitted Parameters	Chemical Processes			
		Keto-enol Tautomerization [7] *Sub* -Arrhenius (Deep-Tunneling)	**OH + H_2 → H + H_2 [76] *Sub*-Arrhenius (Moderate Tunneling)**	**Enzymatic Catalysis [78] *Super*-Arrhenius**	**OH + HBr → Br + H_2O [84] *Anti*-Arrhenius**
Arrhenius $k(T) = A\exp\left(-\frac{E_a}{k_B T}\right)$	A	1.74×10^{3}	$2.16 \cdot 10^{-11}$	1.52×10^{11}	1.66×10^{-11}
	E_a	214	4891	14600	−94.6
	$\chi 2$	1.10×10^{-2}	$4.20 \cdot 10^{-3}$	2.60×10^{-2}	6.69×10^{-2}
Aquilanti–Mundim (AM) $k(T) = A\left(1 - d\frac{\varepsilon^\ddagger}{k_B T}\right)^{\frac{1}{d}}$	A	3.32×10^{6}	$1.11 \cdot 10^{-10}$	1.91×10^{4}	7.43×10^{-14}
	$\varepsilon^\ddagger$	318.06	9170	2391	−324.61
	d	−0.81	−0.086	0.207	1.24
	$\chi 2$	3.68×10^{-2}	$6.80 \cdot 10^{-4}$	2.91×10^{-2}	2.78×10^{-3}
Aquilanti–Sanchez–Coutinho–Carvalho (ASCC) $k(T) = A\left(1 - d\frac{\varepsilon^\ddagger}{k_B T + E_v}\right)^{\frac{1}{d}},\ d = -\frac{1}{3}\left(\frac{E_v}{2\varepsilon^\ddagger}\right)^2$	A	2.33×10^{4}	-	-	-
	$\varepsilon^\ddagger$	2441	-	-	-
	E_v	429	-	-	-
	$\chi 2$	2.18×10^{-2}	-	-	-
Sato–Nakamura–Takayanagi (NTS) $k(T) = A\exp\left[-\frac{E_0}{k_B\left(T^2 + T_0^2\right)^{\frac{1}{2}}}\right]$	A	3.12×10^{4}	-	-	-
	E_0	1655	-	-	-
	T_0	168	-	-	-
	$\chi 2$	7.38×10^{-3}	-	-	-
Vogel–Fulcher–Tammann (VFT) $k(T) = A\exp\left(\frac{B}{T - T_0}\right)$	A	-	-	1.25×10^{5}	-
	B	-	-	−1298	-
	T_0	-	-	175	-
	$\chi 2$	-	-	2.16×10^{-2}	-

Curvature is easily identified through the Arrhenius plot in non-Arrhenius processes; however, the application of the phenomenological formulas to fit this behavior can lead to multiple solutions that make the physical interpretation of the obtained parameters difficult. In Reference [27], we show that curvature in the Arrhenius plot can be linearized using the transitivity plot. Analogous definition to transitivity function were proposed at 1980s to evaluate glass transition of supercooled materials [85–87]. However, the definitions failed to propose a connection with the Tolman theorem, Equation (1), being important only as scaling tools for relaxation processes. Figure 3 shows the Transitivity plot for the temperature dependence of relaxation time of the propylene carbonate [86], where a linear behavior is observed for a certain temperature range as expected by the deformed Aquilanti–Mundim law. However, at 198 K a transition to another linear regime is observed—a break not perceived in the Arrhenius plot. With the linearization of the data in the transitivity plot, the fitting process becomes much simpler: In the first high temperature range the Aquilanti–Mundim parameters are $\frac{1}{\varepsilon^{\ddagger}} = 1.71\ K^{-1}$ and $d = 0.32$; for the second lower temperature range, $\frac{1}{\varepsilon^{\ddagger}} = 0.53K^{-1}$ and $d = 0.08$. The interpretation of these results will be published in a future paper considering a large variety of examples.

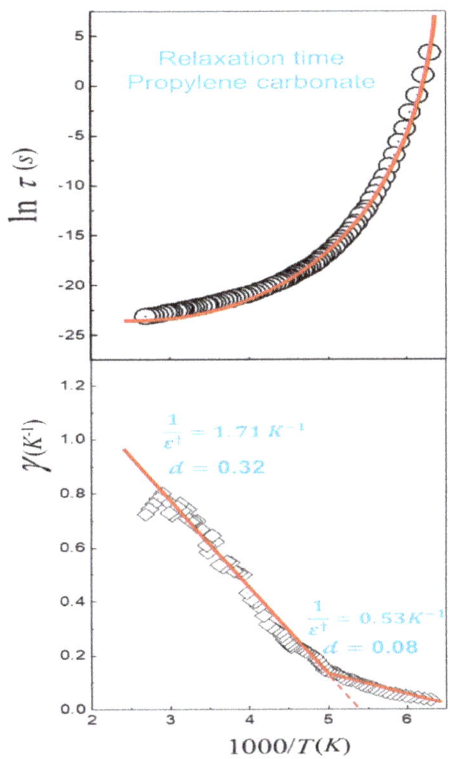

Figure 3. The Arrhenius (upper panel) and Transitivity (lower panel) planes of the temperature dependence of relaxation time of the propylene carbonate. The diamond symbols represent the transitivity values obtained numerically and smoothing with the Savitzky–Golay filter. Red lines emphasize two regions where the temperature dependence of the transitivity is linearized, as expected by the Aquilanti–Mundim law.

4.2. Reaction Rate Constants' Mode

The estimation of the rate constants for the OH + HCl → H_2O + Cl reaction in the gas-phase was performed to validate TST and tunneling corrections, implemented in the Transitivity code.

Furthermore, using the suggestion of the Eyringpy code [54], the $NH_3 + OH \rightarrow NH_2 + H_2O$ reaction was selected to demonstrate the accuracy of the Collins–Kimball and Kramers models to estimate reaction rate constants in an aqueous solution.

4.2.1. The $OH + HCl \rightarrow H_2O + Cl$ Reaction

The reaction rate constants for the reaction between hydroxyl radical and hydrogen chloride are only slightly dependent on temperature in the range 138–300 K, although, as the range increases further, a temperature dependence is observed. This strong concave curvature detected in the Arrhenius plot (sub-Arrhenius behavior) is very convincing evidence of the role of deep quantum tunneling for this reaction [88]. We employed the *d*-TST, Bell35, Bell58, and ST tunneling corrections to calculate the rate constants for the reaction of OH + HCl over a wide range of temperatures (200–2000 K). The electronic structure properties of the reactants, of the products, and of the transition state were calculated employing the MP2/aug-cc-pVDZ calculation level using Gaussian 09 [89]. A complete study of this reaction using our methodology can be found in Reference [44].

Figure 4 obtained by the Transitivity code shows the comparison between the calculated reaction rate constant and the experimental data [90]. No major differences were found using either the Bel58-2T and ST tunneling corrections. Conversely, as expected, a divergence is observed of the Bell58 formula at T_c, and *d*-TST does not describe the range of experimental data at low temperature, where the deep-tunneling regime becomes dominant: This confirms that its validity is limited to weak tunneling. Traditional TST is presented for comparative purposes and visualizes the need for corrections.

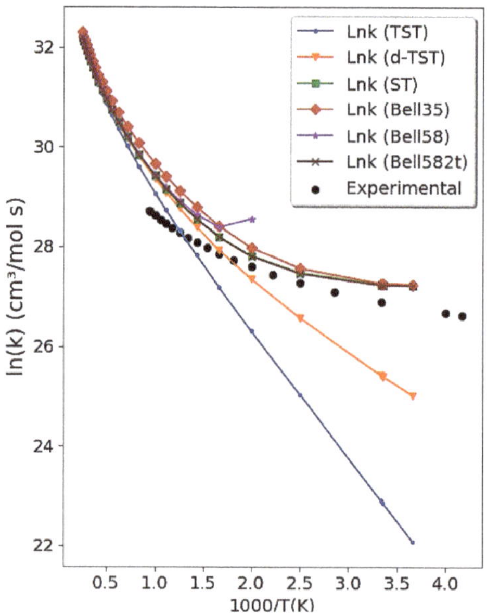

Figure 4. Arrhenius plot obtained from the Transitivity code for the $OH + HCl \rightarrow Cl + H_2O$ reaction using TST with Bell35, Bell58, ST tunneling correction, and *d*-TST. Experimental data in the literature [90] are available for comparison and shown as full dots.

4.2.2. The $NH_3 + OH \rightarrow NH_2 + H_2O$ Reaction

The $NH_3 + OH \rightarrow NH_2 + H_2O$ reaction permits to illustrate the accuracy of the methodology in the liquid-phase. Energies, geometries, and frequencies of stationary points were extracted at the same level of calculation used in the Eyringpy code [54].

The upper panels in Figure 5 show the temperature dependence of the overall reaction rate constant, k_{Obs}, for NH_3 + OH → NH_2 + H_2O reaction from 273 to 4000 K using Kramers and Collins–Kimball models. The Smoluchowski diffusion rate constant $\vec{k}_D$, which evaluates the diffusion limit for a bimolecular reactive process including the solvent effect, is shown in the lower right panel. The Kramers transmission correction, which evaluates the interference of the friction effect of the solvent in the reactive process as a function of temperature, is shown in the lower left panel. At 298.15 K, Kramers' formulation gives for the reaction rate constant the value 6.73×10^{11} cm^3 mol^{-1} s^{-1}, while the Collins-Kimball formulation yields 6.77×10^{11} cm^3 mol^{-1} s^{-1}, (the experimental value indicates ~10^{11} cm^3 mol^{-1} s^{-1}) [91–93]. The value of $\vec{k}_D$ from Smoluchowski (Collins–Kimball) is 3.73×10^{12} cm^3 mol^{-1} s^{-1}, in accordance with that calculated in Reference [54], $\vec{k}_D = 3.60 \times 10^{12}$ cm^3 mol^{-1} s^{-1}.

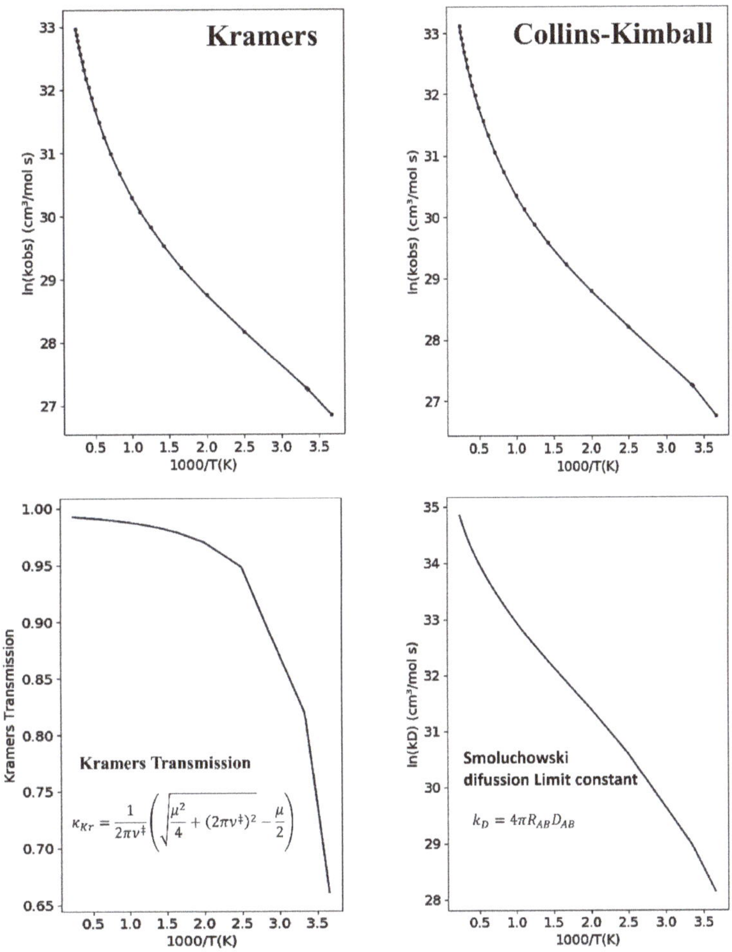

Figure 5. Upper panels present the Arrhenius plots as given by the program for the NH_3 + OH → NH_2 + H_2O reaction using Kramers' and Collins–Kimball formulations. The lower panels show the Kramers transmission and Smoluchowski diffusion limit constant as a function of inverse temperature.

4.3. CPMD Input Files Generator

Coordinates of specific molecules can be selected to test the function of creating a first-principles molecular dynamics input to run by the CPMD code (Figure 6). When selecting this option, the user must choose between the CPMD, PIMD, TSH, MTD and BOMD approaches and must provide the molecular geometry of the system in *.xyz format. Additional information (DFT functional, simulation temperature, system charge, the maximum number of steps, integration time step, and the size of the box) can be inserted through the indicated boxes. Furthermore, it is possible to generate several aleatory configurations by selecting specific initial conditions for the geometric parameters (For more information, see the input files in the Example directory on the www.vhcsgroup.com/transitivity web page).

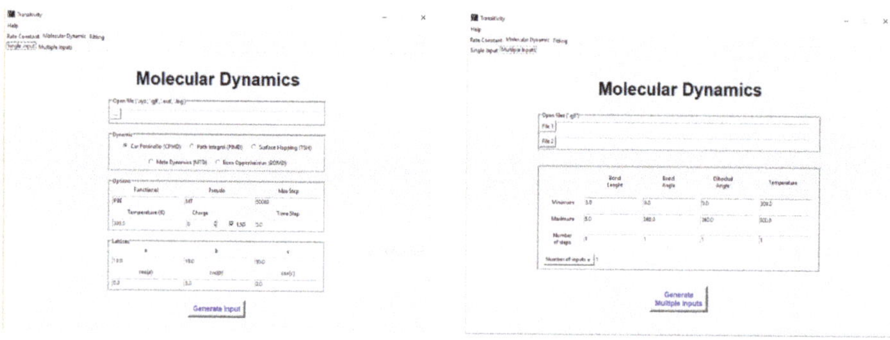

Figure 6. An exemplary view of the input generation function of the transitivity program. Details of input files can be found in the www.vhcsgroup.com/transitivity web page.

5. Final Remarks

The Transitivity code presented in this article tends to systematize the tools developed over the last ten years [25] to handle the kinetics of processes beyond Arrhenius, providing options for: (i) The phenomenological fitting and estimation of reaction rate constants in gas- and liquid-phases and (ii) the preparation of first-principles molecular dynamics in order to evaluate the parameters relative to a variety of reactive processes. In addition, the code provides an easy user-friendly interface and may be relevant for didactic purposes.

A characterizing feature of the code is the consistent use of the *d*-formulation, which recently culminated in a series of successful applications from phenomenological to first-principles descriptions of pure and applied chemical kinetics and material science. Examples are available:

- From the phenomenology of elementary processes (such as the H_2 + F [94], OH + HBr [80], F + HD [65] and C + CH^+ [95] reactions) to complex processes (such as food systems [96], plant respiration [25], plasma chemistry [97], and solid-state diffusive reaction [98]);
- Calculation of the kinetic rate constants for chemical reactions from the potential energy surface features profile, such as the CH_4 + OH [60], CH_3OH + H [99], OH + HCl [44], OH + HI [43], to proton rearrangement of enol forms of curcumin [100], OH + H_2 [101], and chiral nucleophilic substitution reaction [102].

Regarding the evolution and consistency of the code, efforts are being made to introduce *d* predictive formulations to transport properties, as well as post-processing of trajectories obtained by first-principles molecular dynamics simulations. Furthermore, the Eckart tunneling correction and the variational Transition-State Theory represent focuses for the future implementation of the Transitivity code.

Author Contributions: V.H.C.-S. and F.P. conceived and supervised the study. V.H.C.-S. and N.D.C. wrote the paper. H.G.M. and F. O.S.-N. developed the python code, performed stationary electronic structure calculations and fitting procedures. K.C.M. developed and implemented the GSA Fortran code. The manuscript was discussed and reviewed by all authors.

Funding: This research was funded by Brazilian agency CNPq, grant number Universal 01/2016 - Faixa A - 406063/2016-8], by Organizzazione Internazionale Italo-latino Americana (IILA) for a Biotechnology Sector-2019 scholarship and by the Italian Ministry for Education, University and Research, MIUR, for financial support: SIR 2014 "Scientific Independence for Young Researchers" (RBSI14U3VF).

Acknowledgments: The authors are grateful for the support given by Brazilian agency CAPES. This research is also supported by the High-Performance Computing Center at the Universidade Estadual de Goiás, Brazil. Valter H. Carvalho-Silva thanks Brazilian agency CNPq for the research funding programs [Universal 01/2016 - Faixa A - 406063/2016-8] and Organizzazione Internazionale Italo-latino Americana (IILA) for a Biotechnology Sector-2019 scholarship. Federico Palazzetti and Nayara D. Coutinho acknowledge the Italian Ministry for Education, University and Research, MIUR, for financial support: SIR 2014 "Scientific Independence for Young Researchers" (RBSI14U3VF). We thank professor Vincenzo Aquilanti for fruitful discussions.

Conflicts of Interest: The authors declare that the research was conducted in the absence of any commercial or financial relationships that could be construed as a potential conflict of interest.

Appendix A

Table A1. List of Symbols and Nomenclatures.

Symbols	Nomenclature
k	Rate constant
T	Temperature
k_B	Boltzmann constant
β	Lagrange multiplier
γ	Transitivity function
d	Deformed parameter
h	Planck's constant
Q	Partition functions
AM	Aquilanti-Mundim
dH	Enthalpy of reaction
ASCC	Aquilanti–Sanchez–Coutinho–Carvalho
NTS	Nakamura–Takayanagi–Sato
TST	Transition-State Theory
GSA	Generalized Simulated Annealing
d-TST	*Deformed* Transition-State Theory
ST	Skodje and Truhlar tunneling correction
Bell35	Bell's tunneling correction of 1935
Bell58	Bell's tunneling correction of 1958
$\varepsilon^{\ddagger}$	barrier height (Eyring's parameter)
E_a	Apparent Activation Energy
E_0	Energy parameter from NTS formula
E_v	Energy parameter from ASCC formula
B	Temperature parameter from VFT formula
T_0	Temperature parameter from NTS and VFT formulas.
T_c	Crossover temperature
$\vec{k}_D$	Diffusion rate constant
$\nu^{\ddagger}$	Imaginary frequency
k_{Obs}	Overall reaction rate constant
κ_{Kr}	Transmission factor from Kramers' model
μ	Friction constant
η	Viscosity
DFT	Density functional theory
BOMD	Born-Oppenheimer molecular dynamics
CPMD	Car-Parrinello molecular dynamics
PIMD	Path-Integral molecular dynamics
MTD	Metadynamics
TSH	Trajectory Surface Hopping

References

1. Valter Henrique, C.-S.; Coutinho, N.D.; Aquilanti, V. From the Kinetic Theory of Gases to the Kinetics of Chemical Reactions: On the Verge of the Thermodynamical and the Kinetic Limits. *Molecules* **2019**. to be submitted.
2. Aquilanti, V.; Coutinho, N.D.; Carvalho-Silva, V.H. Kinetics of Low-Temperature Transitions and Reaction Rate Theory from Non-Equilibrium Distributions. *Philos. Trans. R. Soc. London A* **2017**, *375*, 20160204. [CrossRef]
3. Aquilanti, V.; Borges, E.P.; Coutinho, N.D.; Mundim, K.C.; Carvalho-Silva, V.H. From statistical thermodynamics to molecular kinetics: the change, the chance and the choice. *Rend. Lincei. Sci. Fis. e Nat.* **2018**, *28*, 787–802. [CrossRef]
4. Gentili, P.L. The fuzziness of the molecular world and its perspectives. *Molecules* **2018**, *23*, 2074. [CrossRef]
5. Gentili, P.L. *Untangling Complex Systems: A Grand Challenge for Science*; CRC Press: Boca Raton, FL, USA, 2018; ISBN 9780429847547.
6. Atkinson, R. Kinetics and mechanisms of the gas-phase reactions of the hydroxyl radical with organic compounds under atmospheric conditions. *Chem. Rev.* **1986**, *86*, 69–201. [CrossRef]
7. Limbach, H.-H.; Miguel Lopez, J.; Kohen, A. Arrhenius curves of hydrogen transfers: tunnel effects, isotope effects and effects of pre-equilibria. *Philos. Trans. R. Soc. B Biol. Sci.* **2006**, *361*, 1399–1415. [CrossRef]
8. Smith, I.W.M. The temperature-dependence of elementary reaction rates: Beyond Arrhenius. *Chem. Soc. Rev.* **2008**, *37*, 812–826. [CrossRef]
9. Sims, I.R. Low-temperature reactions: Tunnelling in space. *Nat. Chem.* **2013**, *5*, 734–736. [CrossRef] [PubMed]
10. Peleg, M.; Normand, M.D.; Corradini, M.G. The Arrhenius Equation Revisited. *Crit. Rev. Food Sci. Nutr.* **2012**, *52*, 830–851. [CrossRef] [PubMed]
11. Darrington, R.T.; Jiao, J. Rapid and Accurate Prediction of Degradant Formation Rates in Pharmaceutical Formulations Using High-Performance Liquid Chromatography-Mass Spectrometry. *J. Pharm. Sci.* **2004**, *93*, 838–846. [CrossRef]
12. Giordano, D.; Russell, J.K. Towards a structural model for the viscosity of geological melts. *Earth Planet. Sci. Lett.* **2018**, *501*, 202–212. [CrossRef]
13. Klinman, J.P.; Kohen, A. Hydrogen Tunneling Links Protein Dynamics to Enzyme Catalysis. *Annu. Rev. Biochem.* **2013**, *82*, 471–496. [CrossRef]
14. Warshel, A.; Bora, R.P. Perspective: Defining and quantifying the role of dynamics in enzyme catalysis. *J. Chem. Phys.* **2016**, *144*, 180901. [CrossRef]
15. Laidler, K.J. A Glossary of Terms Used in Chemical Kinetics, Including Reaction Dynamics. *Pure Appl. Chem.* **1996**, *68*, 149–192. [CrossRef]
16. Tolman, R.C. Statistical Mechanics Applied to Chemical Kinetics. *J. Amer. Chem. Soc.* **1920**, *42*, 2506–2528. [CrossRef]
17. Glasstone, S.; Laidler, K.J.; Eyring, H. *The Theory of Rate Processes: The Kinetics of Chemical Reactions, Viscosity, Diffusion and Electrochemical Phenomena*; International chemical series; McGraw-Hill: New York, NY, USA, 1941.
18. Truhlar, D.G. Current Status of Transition-State Theory. *J. Phys. Chem.* **1983**, 2664–2682. [CrossRef]
19. Kooij, D.M. Über die Zersetzung des gasförmigen Phosphorwasserstoffs. *Zeitschrift für Phys. Chemie* **1893**, *12*, 155–161.
20. Bělehrádek, J. A unified theory of cellular rate processes based upon an analysis of temperature action. *Protoplasma* **1957**, *48*, 53–71. [CrossRef]
21. Vogel, H. Das temperature-abhangigketsgesetz der viskositat von flussigkeiten. *Phys. Z* **1921**, *22*, 645–646.
22. Fulcher, G.S. Analysis of Recent Measurements of the Viscosity of Glasses. *J. Am. Ceram. Soc.* **1925**, *8*, 339–355. [CrossRef]
23. Tammann, G.; Hesse, W. Die Abhängigkeit der Viscosität von der Temperatur bie unterkühlten Flüssigkeiten. *Zeitschrift für Anorg. und Allg. Chemie* **1926**, *156*, 245–257. [CrossRef]
24. Nakamura, K.; Takayanagi, T.; Sato, S. A modified arrhenius equation. *Chem. Phys. Lett.* **1989**, *160*, 295–298. [CrossRef]

25. Aquilanti, V.; Mundim, K.C.; Elango, M.; Kleijn, S.; Kasai, T. Temperature dependence of chemical and biophysical rate processes: Phenomenological approach to deviations from Arrhenius law. *Chem. Phys. Lett.* **2010**, *498*, 209–213. [CrossRef]
26. Coutinho, N.D.; Silva, Y.S.; de Fazio, D.; Cavalli, S.; Carvalho-Silva, V.H.; Aquilanti, V. Chemical Kinetics under Extreme Conditions: Exact, Phenomenological and First-Principles Computational Approaches. In *The Astrochemical Observatory: Focus on Chiral Molecules*; Accademia Nazionale delle Scienze detta dei XL: Rome, Italy, 2018; pp. 1–15.
27. Carvalho-Silva, V.H.; Coutinho, N.D.; Aquilanti, V. Temperature dependence of rate processes beyond Arrhenius and Eyring: Activation and Transitivity. *Front. Chem.* **2019**, *7*, 380. [CrossRef] [PubMed]
28. Fernandez-Ramos, A.; Ellingson, B.A.; Garrett, B.C.; Truhlar, D.G. Variational Transition State Theory with Multidimensional Tunneling. In *Reviews in Computational Chemistry*; John Wiley & Sons, Inc.: Hoboken, NJ, USA, 2007; pp. 125–232. ISBN 9780470116449.
29. Marcus, R.A. Electron transfer reactions in chemistry. Theory and experiment. *Rev. Mod. Phys.* **1993**, *65*, 599–610. [CrossRef]
30. Miller, W. Quantum mechanical transition state theory and a new semiclassical model for reaction rate constants. *J. Chem. Phys.* **1974**, *61*, 1823. [CrossRef]
31. Richardson, J.O. Ring-Polymer Approaches to Instanton Theory. Ph.D. Thesis, University of Cambridge, Cambridge, UK, 2012.
32. Zhang, Y.; Stecher, T.; Cvitaš, M.T.; Althorpe, S.C. Which Is Better at Predicting Quantum-Tunneling Rates: Quantum Transition-State Theory or Free-Energy Instanton Theory? *J. Phys. Chem. Lett.* **2014**, *5*, 3976–3980. [CrossRef]
33. Xiao, R.; Gao, L.; Wei, Z.; Spinney, R.; Luo, S.; Wang, D.; Dionysiou, D.D.; Tang, C.J.; Yang, W. Mechanistic insight into degradation of endocrine disrupting chemical by hydroxyl radical: An experimental and theoretical approach. *Environ. Pollut.* **2017**, *231*, 1446–1452. [CrossRef]
34. Luo, S.; Gao, L.; Wei, Z.; Spinney, R.; Dionysiou, D.D.; Hu, W.P.; Chai, L.; Xiao, R. Kinetic and mechanistic aspects of hydroxyl radical-mediated degradation of naproxen and reaction intermediates. *Water Res.* **2018**, *137*, 233–241. [CrossRef]
35. Gao, Y.; Ji, Y.; Li, G.; An, T. Mechanism, kinetics and toxicity assessment of OH-initiated transformation of triclosan in aquatic environments. *Water Res.* **2014**, *49*, 360–370. [CrossRef]
36. De Sainte Claire, P. Degradation of PEO in the solid state: A theoretical kinetic model. *Macromolecules* **2009**, *42*, 3469–3482. [CrossRef]
37. Ahubelem, N.; Shah, K.; Moghtaderi, B.; Page, A.J. Formation of benzofuran and chlorobenzofuran from 1,3-dichloropropene: A quantum chemical investigation. *Int. J. Quantum Chem.* **2015**, *115*, 1739–1745. [CrossRef]
38. Zavala-Oseguera, C.; Galano, A.; Merino, G. Computational study on the kinetics and mechanism of the carbaryl + OH reaction. *J. Phys. Chem. A* **2014**, *118*, 7776–7781. [CrossRef] [PubMed]
39. Döntgen, M.; Przybylski-Freund, M.-D.; Kröger, L.C.; Kopp, W.A.; Ismail, A.E.; Leonhard, K. Automated discovery of reaction pathways, rate constants, and transition states using reactive molecular dynamics simulations. *J. Chem. Theory Comput.* **2015**, *11*, 2517–2524. [CrossRef] [PubMed]
40. Piccini, G.; McCarty, J.; Valsson, O.; Parrinell, M. Variational Flooding Study of a S2 Reaction. *J. Phys. Chem. A* **2017**, *8*, 580–583.
41. Fleming, K.L.; Tiwary, P.; Pfaendtner, J. New Approach for Investigating Reaction Dynamics and Rates with Ab Initio Calculations. *J. Phys. Chem. A* **2016**, *120*, 299–305. [CrossRef] [PubMed]
42. Lancar, I.T.; Mellouki, A.; Poulet, G. Kinetics of the reactions of hydrogen iodide with hydroxyl and nitrate radicals. *Chem. Phys. Lett.* **1991**, *177*, 554–558. [CrossRef]
43. Coutinho, N.D.; Carvalho-Silva, V.H.; de Oliveira, H.C.B.; Aquilanti, V. *The HI + OH → H2O + I Reaction by First-Principles Molecular Dynamics: Stereodirectional and Anti-Arrhenius Kinetics*, 2017; Volume 10408 LNCS, ISBN 9783319624037.
44. Coutinho, N.D.; Sanches-Neto, F.O.; Carvalho-Silva, V.H.; de Oliveira, H.C.B.; Ribeiro, L.A.; Aquilanti, V. Kinetics of the OH+HCl→H2O+Cl reaction: Rate determining roles of stereodynamics and roaming and of quantum tunneling. *J. Comput. Chem.* **2018**, *39*, 2508–2516. [CrossRef] [PubMed]

45. Isaacson, A.D.; Truhlar, D.G.; Rai, S.N.; Steckler, R.; Hancock, G.C.; Garrett, B.C.; Redmon, M.J. POLYRATE: A general computer program for variational transition state theory and semiclassical tunneling calculations of chemical reaction rates. *Comput. Phys. Commun.* **1987**, *47*, 91–102. [CrossRef]
46. Duncan, W.T.; Bell, R.L.; Truong, T.N. TheRate: Program forab initio direct dynamics calculations of thermal and vibrational-state-selected rate constants. *J. Comput. Chem.* **1998**, *19*, 1039–1052. [CrossRef]
47. Barker, J.R. Multiple-Well, multiple-path unimolecular reaction systems. I. MultiWell computer program suite. *Int. J. Chem. Kinet.* **2001**, *33*, 232–245. [CrossRef]
48. Ghysels, A.; Verstraelen, T.; Hemelsoet, K.; Waroquier, M.; Van Speybroeck, V. TAMkin: A versatile package for vibrational analysis and chemical kinetics. *J. Chem. Inf. Model.* **2010**, *50*, 1736–1750. [CrossRef] [PubMed]
49. Glowacki, D.R.; Liang, C.-H.; Morley, C.; Pilling, M.J.; Robertson, S.H. MESMER: An Open-Source Master Equation Solver for Multi-Energy Well Reactions. *J. Phys. Chem. A* **2012**, *116*, 9545–9560. [CrossRef] [PubMed]
50. Gao, C.W.; Allen, J.W.; Green, W.H.; West, R.H. Reaction Mechanism Generator: Automatic construction of chemical kinetic mechanisms. *Comput. Phys. Commun.* **2016**, *203*, 212–225. [CrossRef]
51. Euclides, H.O.; Barreto, P.R. APUAMA: a software tool for reaction rate calculations. *J. Mol. Model.* **2017**, *23*, 176. [CrossRef] [PubMed]
52. Canneaux, S.; Bohr, F.; Henon, E. KiSThelP: A program to predict thermodynamic properties and rate constants from quantum chemistry results. *J. Comput. Chem.* **2014**, *35*, 82–93. [CrossRef] [PubMed]
53. Coppola, C.M. Mher V Kazandjian; Matrix formulation of the energy exchange problem of multi-level systems and the code FRIGUS. *Rend. Lincei Sci. Fis. e Nat.* **2019**, in press.
54. Dzib, E.; Cabellos, J.L.; Ortíz-Chi, F.; Pan, S.; Galano, A.; Merino, G. Eyringpy: A program for computing rate constants in the gas phase and in solution. *Int. J. Quantum Chem.* **2018**, *119*, 11–13. [CrossRef]
55. Mundim, K.C.; Tsallis, C. Geometry optimization and conformational analysis through generalized simulated annealing. *Int. J. Quantum Chem.* **1998**, *58*, 373–381. [CrossRef]
56. Sato, S. Tunneling in bimolecular reactions. *Chem. Phys.* **2005**, *315*, 65–75. [CrossRef]
57. Bell, R.P. Quantum Mechanical Effects in Reactions Involving Hydrogen. *Proc. R. Soc. London. Ser. A, Math. Phys. Sci.* **1935**, *CXLVIII.A*, 241–250.
58. Bell, R.P. The Tunnel Effect Correction For Parabolic Potential Barriers. *Faraday Soc. Contrib.* **1958**, 1–4. [CrossRef]
59. Skodje, R.T.; Truhlar, D.G. Parabolic tunneling calculations. *J. Phys. Chem.* **1981**, *85*, 624–628. [CrossRef]
60. Carvalho-Silva, V.H.; Aquilanti, V.; de Oliveira, H.C.B.; Mundim, K.C. Deformed transition-state theory: Deviation from Arrhenius behavior and application to bimolecular hydrogen transfer reaction rates in the tunneling regime. *J. Comput. Chem.* **2017**, *38*, 178–188. [CrossRef] [PubMed]
61. Collins, F.C.; Kimball, G.E. Diffusion-controlled reaction rates. *J. Colloid Sci.* **1949**, *4*, 425–437. [CrossRef]
62. Kramers, H.A. Brownian motion in a field of force and the diffusion model of chemical reactions. *Phys.* **1940**, *7*, 284–304. [CrossRef]
63. CPMDversion 3.17.1; CPMD, version 4.; CPMDversion 4.1; CPMDversion 3.17.1 Copyright IBM 2012.
64. Claudino, D.; Gargano, R.; Carvalho-Silva, V.H.; E Silva, G.M.; Da Cunha, W.F. Investigation of the Abstraction and Dissociation Mechanism in the Nitrogen Trifluoride Channels: Combined Post-Hartree-Fock and Transition State Theory Approaches. *J. Phys. Chem. A* **2016**, *120*, 5464–5473. [CrossRef]
65. Cavalli, S.; Aquilanti, V.; Mundim, K.C.; De Fazio, D. Theoretical reaction kinetics astride the transition between moderate and deep tunneling regimes: The F + HD case. *J. Phys. Chem. A* **2014**, *118*, 6632–6641. [CrossRef]
66. Bell, R.P. *The Tunnel Effect in Chemistry*; Champman and Hall: London, UK, 1980.
67. Christov, S.G. The Characteristic (Crossover) Temperature in the Theory of Thermally Activated Tunneling Processes. *Mol. Eng.* **1997**, *7*, 109–147. [CrossRef]
68. Onsager, L. Electric Moments of Molecules in Liquids. *J. Am. Chem. Soc.* **1936**, *58*, 1486–1493. [CrossRef]
69. Wong, M.W.; Wiberg, K.B.; Frisch, M.J. Solvent effects. 3. Tautomeric equilibria of formamide and 2-pyridone in the gas phase and solution: an ab initio SCRF study. *J. Am. Chem. Soc.* **1992**, *114*, 1645–1652. [CrossRef]
70. Henriksen, N.E.; Hansen, F.Y. *Theories of Molecular Reaction Dynamics: The Microscopic Foundation of Chemical Kinetics*; Oxford University Press: New York, NY, USA, 2008; ISBN 9780191708251.
71. Smoluchowski, M.V. Drei Vortrage uber Diffusion, Brownsche Bewegung und Koagulation von Kolloidteilchen. *Phys. Zeit.* **1916**, *17*, 557–585.

72. Collins, F.C.; Kimball, G.E. Diffusion-Controlled Reactions in Liquid Solutions. *Ind. Eng. Chem.* **1949**, *41*, 2551–2553. [CrossRef]
73. Eigen, M. Acid-Base Catalysis, and Enzymatic Hydrolysis. Part I: Elementary Processes. *Angew. Chemie Int. Ed. English* **1964**, *3*, 1–19. [CrossRef]
74. Hallett, J. The Temperature Dependence of the Viscosity of Supercooled Water. *Proc. Phys. Soc.* **1963**, *82*, 1046–1050. [CrossRef]
75. Savitzky, A.; Golay, M.J.E. Smoothing and Differentiation of Data by Simplified Least Squares Procedures. *Anal. Chem.* **1964**, *36*, 1627–1639. [CrossRef]
76. Ravishankara, A.R.; Nicovich, J.M.; Thompson, R.L.; Tully, F.P. Kinetic study of the reaction of hydroxyl with hydrogen and deuterium from 250 to 1050 K. *J. Phys. Chem.* **1981**, *85*, 2498–2503. [CrossRef]
77. Kohen, A.; Cannio, R.; Bartolucci, S.; Klinman, J.P. Enzyme dynamics and hydrogen tunnelling in a thermophilic alcohol dehydrogenase. *Nature* **1999**, *399*, 496–499. [CrossRef]
78. Liang, Z.X.; Tsigos, I.; Bouriotis, V.; Klinman, J.P. Impact of protein flexibility on hydride-transfer parameters in thermophilic and psychrophilic alcohol dehydrogenases. *J. Am. Chem. Soc.* **2004**, *126*, 9500–9501. [CrossRef]
79. Truhlar, D.; Kohen, A. Convex Arrhenius plots and their interpretation. *Proc. Nat. Acad. Sci. USA* **2001**, *98*, 848–851. [CrossRef]
80. Coutinho, N.D.; Aquilanti, V.; Silva, V.H.C.; Camargo, A.J.; Mundim, K.C.; De Oliveira, H.C.B. Stereodirectional Origin of anti-Arrhenius Kinetics for a Tetraatomic Hydrogen Exchange Reaction: Born-Oppenheimer Molecular Dynamics for OH + HBr. *J. Phys. Chem. A* **2016**, *120*, 5408–5417. [CrossRef] [PubMed]
81. de Oliveira-Filho, A.G.S.; Ornellas, F.R.; Bowman, J.M. Quasiclassical Trajectory Calculations of the Rate Constant of the OH + HBr → Br + H_2O Reaction Using a Full-Dimensional Ab Initio Potential Energy Surface Over the Temperature Range 5 to 500 K. *J. Phys. Chem. Lett.* **2014**, *5*, 706–712. [CrossRef] [PubMed]
82. Coutinho, N.D.; Silva, V.H.C.; de Oliveira, H.C.B.; Camargo, A.J.; Mundim, K.C.; Aquilanti, V. Stereodynamical Origin of Anti-Arrhenius Kinetics: Negative Activation Energy and Roaming for a Four-Atom Reaction. *J. Phys. Chem. Lett.* **2015**, *6*, 1553–1558. [CrossRef] [PubMed]
83. Zuniga-Hansen, N.; Silbert, L.E.; Calbi, M.M. Breakdown of kinetic compensation effect in physical desorption. *Phys. Rev. E* **2018**, *98*, 032128. [CrossRef]
84. Sims, I.R.; Smith, I.W.M.; Clary, D.C.; Bocherel, P.; Rowe, B.R. Ultra-low Temperature Kinetics of Neutral-neutral Reactions - New Experimental and Theoretical Results For OH + HBr Between 295 K and 23 K. *J. Chem. Phys.* **1994**, *101*, 1748–1751. [CrossRef]
85. Souletie, J.; Tholence, J.L. Critical slowing down in spin glasses and other glasses: Fulcher versus power law. *Phys. Rev. B* **1985**, *32*, 516. [CrossRef]
86. Stickel, F.; Fischer, E.W.; Richert, R. Dynamics of glass-forming liquids. II. Detailed comparison of dielectric relaxation, de-conductivity, and viscosity data. *J. Chem. Phys.* **1996**, *104*, 2043. [CrossRef]
87. Drozd-Rzoska, A. Universal behavior of the apparent fragility in ultraslow glass forming systems. *Sci. Rep.* **2019**, *9*, 6816. [CrossRef]
88. Silva, V.H.C.; Aquilanti, V.; de Oliveira, H.C.B.; Mundim, K.C. Uniform description of non-Arrhenius temperature dependence of reaction rates, and a heuristic criterion for quantum tunneling vs classical non-extensive distribution. *Chem. Phys. Lett.* **2013**, *590*, 201–207. [CrossRef]
89. Frisch, M.J.; Trucks, G.W.; Schlegel, H.B.; Scuseria, G.E.; Robb, M.A.; Cheeseman, J.R.; Scalmani, G.; Barone, V.; Mennucci, B.; Petersson, G.A.; et al. *Gaussian09 Revision D.01 2009*; Gaussian, Inc.: Wallingford, CT, USA, 2009.
90. Ravishankara, A.R.; Wine, P.H.; Wells, J.R.; Thompson, R.L. Kinetic study of the reaction of OH with HCl from 240 to 1055 K. *Chem. Phys. Lett.* **1985**, *17*, 1281–1297. [CrossRef]
91. Hickel, B.; Sehested, K. Reaction of hydroxyl radicals with ammonia in liquid water at elevated temperatures. *Int. J. Radiat. Appl. Instrum. Part C. Radiat. Phys. Chem.* **1992**, *39*, 355–357. [CrossRef]
92. Men'kin, V.B.; Makarov, I.E.; Pikaev, A.K. Pulse radiolysis study of reaction rates of OH and O-radicals with ammonia in aqueous solutions. *High Energy Chem. (Engl. Transl.)* **1989**, *22*, 333–336.
93. Neta, P.; Maruthamuthu, P.; Carton, P.M.; Fessenden, R.W. Formation and reactivity of the amino radical. *J. Phys. Chem.* **1978**, *82*, 1875–1878. [CrossRef]

94. Aquilanti, V.; Mundim, K.C.; Cavalli, S.; De Fazio, D.; Aguilar, A.; Lucas, J.M. Exact activation energies and phenomenological description of quantum tunneling for model potential energy surfaces. The F+H_2 reaction at low temperature. *Chem. Phys.* **2012**, *398*, 186–191. [CrossRef]
95. Rampino, S.; Pastore, M.; Garcia, E.; Pacifici, L.; Laganà, A. On the temperature dependence of the rate coefficient of formation of C_2^+ from C + CH^+. *Mon. Not. R. Astron. Soc.* **2016**, *460*, 2368–2375. [CrossRef]
96. Coutinho, N.D.; Silva, V.H.C.; Mundim, K.C.; de Oliveira, H.C.B. Description of the effect of temperature on food systems using the deformed Arrhenius rate law: deviations from linearity in logarithmic plots vs. inverse temperature. *Rend. Lincei* **2015**, *26*, 141–149. [CrossRef]
97. Capitelli, M.; Pietanza, L.D. Past and present aspects of Italian plasma chemistry. *Rend. Lincei. Sci. Fis. e Nat.* **2019**. [CrossRef]
98. Agreda, N.J.L. Aquilanti–Mundim deformed Arrhenius model in solid-state reactions: Theoretical evaluation using DSC experimental data. *J. Therm. Anal. Calorim.* **2016**, *126*, 1175–1184. [CrossRef]
99. Sanches-Neto, F.O.; Coutinho, N.D.; Silva, V. A novel assessment of the role of the methyl radical and water formation channel in the CH3OH + H reaction. *Phys. Chem. Chem. Phys.* **2017**, *19*, 24467–24477. [CrossRef] [PubMed]
100. Santin, L.G.; Toledo, E.M.; Carvalho-Silva, V.H.; Camargo, A.J.; Gargano, R.; Oliveira, S.S. Methanol Solvation Effect on the Proton Rearrangement of Curcumin's Enol Forms: An Ab Initio Molecular Dynamics and Electronic Structure Viewpoint. *J. Phys. Chem. C* **2016**, *120*, 19923–19931. [CrossRef]
101. Carvalho-Silva, V.H.; Vaz, E.C.; Coutinho, N.D.; Kobayashi, H.; Kobayashi, Y.; Kasai, T.; Palazzetti, F.; Lombardi, A.; Aquilanti, V. The Increase of the Reactivity of Molecular Hydrogen with Hydroxyl Radical from the Gas Phase versus an Aqueous Environment: Quantum Chemistry and Transition State-Theory Calculations. In *Lecture Notes in Computer Science*; Springer: Cham, Switzerland, 2019; pp. 450–459.
102. Rezende, M.V.C.S.; Coutinho, N.D.; Palazzetti, F.; Lombardi, A.; Carvalho-Silva, V.H. Nucleophilic substitution vs elimination reaction of bisulfide ions with substituted methanes: exploration of chiral selectivity by stereodirectional first-principles dynamics and transition state theory. *J. Mol. Model.* **2019**, *25*, 227. [CrossRef] [PubMed]

Sample Availability: Not available.

MDPI
St. Alban-Anlage 66
4052 Basel
Switzerland
Tel. +41 61 683 77 34
Fax +41 61 302 89 18
www.mdpi.com

Molecules Editorial Office
E-mail: molecules@mdpi.com
www.mdpi.com/journal/molecules

www.ingramcontent.com/pod-product-compliance
Lightning Source LLC
Chambersburg PA
CBHW062044220926
51850CB00010B/2661

* 9 7 8 3 0 3 9 4 3 1 7 8 6 *